SOUND SCIENCE

Also by Etta Kaner and Louise Phillips:

Balloon Science

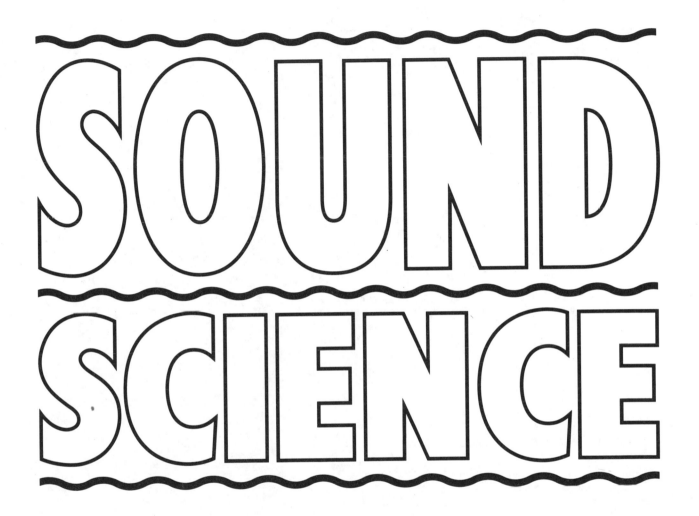

SOUND SCIENCE

Written by Etta Kaner
Illustrated by Louise Phillips

Addison-Wesley Publishing Company
Reading, Massachusetts   Menlo Park, California   New York
Don Mills, Ontario   Wokingham, England   Amsterdam   Bonn
Sydney   Singapore   Tokyo   Madrid   San Juan
Paris   Seoul   Milan   Mexico City   Taipei

Library of Congress Cataloging-in-Publication Data

Kaner, Etta.
  Sound science / written by Etta Kaner ; illustrated by Louise
Phillips.
    p.   cm.
  Includes index.
  Summary: Explores the nature of sound through experiments,
riddles, interesting facts, puzzles, and games.
  ISBN 0-201-56758-X
  1. Sound — Juvenile literature. 2. Sound — Experiments — Juvenile
literature. [1. Sound. 2. Sound — Experiments. 3. Experiments.]
I. Phillips, Louise, ill.   II. Title.
QC225.5.K33   1991
534 — dc20   91-28590
                                                        CIP
                                                         AC

Originally published in Canada by Kids Can Press, Ltd., of Toronto.

Design by Louise Phillips
Cover by Fred Harsh
Edited by Laurie Wark
Set in 12-point ITC Century Book by First Image

Neither the Publisher nor the Author shall be liable for any damage which may
be caused or sustained as a result of conducting any of the activities in this
book without specifically following instructions, conducting the activities without proper supervision, or ignoring the cautions contained in the book.

6 7 8 9-CRS-0099989796
Sixth printing, July 1996

Addison-Wesley books are available at special discounts for bulk purchases by
schools, institutions, and other organizations. For more information, please
contact:

Special Markets Department
Addison-Wesley Publishing Company
Reading, MA 01867
(617) 944-3700 x2431

 Text stock contains over 50% recycled paper

CONTENTS

For Esther and the late Albert Nitkin

Acknowledgements

I want to thank the many people who gave willingly and generously of their time and knowledge during the writing of this book. Without them this book would have been impossible. Thank you to Mel Dwosh for his physics tips, Pat Gagne of the Canadian Pacific Railroad for detailed and fascinating information about train whistles and moose, Malcolm Murray of the Metro Toronto Police for his clear and patient explanation of radar guns, Dael E. Morris of the Royal Ontario Museum for her research on crickets and corn earworms, Dr. J. Pitre of the University of Toronto physics department for his physics explanations, Peter Ryan of the Canadian National Institute for the Blind for his extensive information about the use of sound by blind people, Cathie Spencer of the Ontario Science Centre for her thorough reading of the manuscript and her excellent suggestions and pleasant manner, Dr. Patrick Tevlin of the Ontario Science Centre for his patient sound explanations, Frank Thompson of Brock Public School for his invaluable assistance in setting up the telegraph, Anne–Marie Van Nest of the Civic Garden Centre for her information about music and plants, Ted Venema of the Canadian Hearing Society for his ideas on hearing aids and bone conduction, Jiri Vondrak of Cinram for his detailed explanation of compact discs, Rob Yale of Digital Music Studios for his patient explanation of synthesizers. Special thanks to Louise Phillips who did double duty as designer and illustrator; Laurie Wark, my editor at Kids Can Press, for her patience, attention to detail and good humour; and to my wonderful family who helped me with the many experiments and improved on my riddles.

INTRODUCTION

What can you make but not see? What can travel through solids but make no holes in them? Give up? It's sound. In this book, you'll find out how to do all this and more when you experiment with sound. You'll find out how to change the pitch of sounds, how to amplify, or make sound louder, and how sound travels through water, air, wood, steel and bones. You can grow plants with sound or make your own musical instruments. Investigate Morse code by making a telegraph, or experiment with your own hearing, sound effects, echoes, ventriloquism and much more. You'll discover some interesting facts about body noises, animal sounds, world records, noise pollution and famous inventors. You'll tickle your funny bone with silly riddles, and you can challenge yourself and your friends with sound puzzles and games. Here's one challenge to get you started. How many words can you find in this book that have the word "phone" in them? "Phone" means sound in Greek. Happy hunting and happy experimenting!

If you find a sound word that you don't understand, check the glossary on page 95 for an explanation.

I'M ALL EARS

Whether your ears stick out from the sides of your head
or are more laid back, they do the same thing.
They funnel sound into the inner part of your ear. But are they
the best possible shape for doing this? And what happens inside your
ear once sound enters it? Are there other ways that people can hear?
Find the answers to these questions when you try
the experiments in this section.

A LOOK INSIDE YOUR EAR

If you could look inside your ear,
you would see a thin membrane about the width
of a pea stretched across the end of a short tube.
This is your eardrum. When sounds reach your ear, your eardrum
sets in motion an amazing arrangement of tiny bones, tubes, hairs
and nerves that works together with your brain to let you hear sounds.
Try this experiment to see how your ear works.

You'll need:
- *an empty frozen juice can*
- *a can opener*
- *a balloon*
- *a rubber band*
- *glue*
- *a piece of mirror that is 1/2 cm (1/4 inch) square (Ask a mirror store to cut off a piece from their scraps.)*
- *a dark room*
- *a flashlight*

2 Inflate the balloon and hold it closed for a minute to stretch the rubber. Let out the air. Cut the balloon in half across its width. Discard the neck half.

1 Use a can opener to remove both ends of the can.

3 Stretch the balloon over one end of the can so that the balloon is very taut. If necessary, hold it in place with a rubber band.

4 Glue the mirror to the outside of the stretched balloon about 1 cm (1/2 inch) from the edge of the can.

5 In a dark room, shine the flashlight at an angle onto the mirror so that a bright spot is reflected from the mirror onto the wall or ceiling.

6 Shout into the can from the open end. What happens to the spot on the wall? Now make softer sounds and sing high and low notes. What do you see?

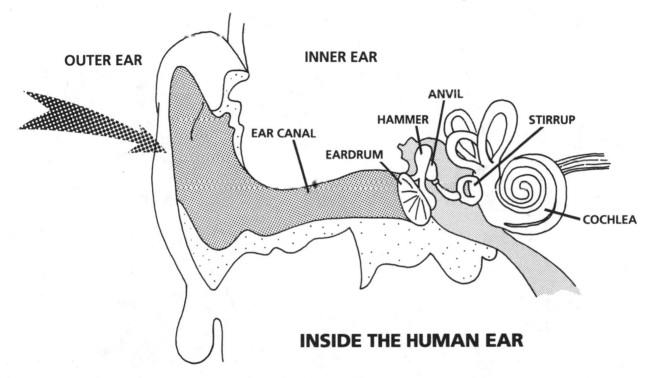

OUTER EAR

INNER EAR

EAR CANAL

EARDRUM

HAMMER

ANVIL

STIRRUP

COCHLEA

INSIDE THE HUMAN EAR

❓ How does it work?

When you shout into the can, the sound travels through the air in the can to the stretched balloon. The sound makes the balloon vibrate, which makes the mirror vibrate. You can see the vibrations when you look at the mirror's reflection on the wall.

A similar thing happens inside your ear when you hear a sound. The sound is collected by your outer ear and travels down a small tube, called the ear canal, to your eardrum. When sound reaches your eardrum, it vibrates just like the balloon did.

The vibrating eardrum pushes against three tiny bones: the hammer, anvil and stirrup. These bones make the vibrations louder. The stirrup then pushes against a coiled tube filled with liquid. The tube, or cochlea, is lined with thousands of tiny hairs. When sound vibrations make the liquid move, the hairs wave back and forth. These moving hairs are connected to nerves that carry the sound message to your brain. All this happens in a split second.

DO WE REALLY NEED BOTH EARS?

Cover up one ear. Can you still hear? Sure you can.
Then why do you need the other ear?
Here's an experiment to help you solve this mystery.

You'll need:
- *a scarf or something to use as a blindfold*
- *a friend*
- *sticky tape*
- *2 pennies*

1 Blindfold your friend, making sure that her ears are uncovered. Have her sit in the middle of a quiet room without moving her head.

2 Make a small strip of tape into a circle with the sticky side on the outside. Stick one side of the tape onto a penny and the other side onto the pad of your thumb.

3 Tape the other penny onto your forefinger of the same hand in the same way.

4 Tiptoe to different parts of the room and click the pennies twice in each location. Have your friend point in the direction of each set of clicks. Keep a record of how many times she guesses right and wrong.

5 Now have your friend press the palm of one hand against one ear. Repeat the above experiment the same number of times. How do the two records compare?

? How does it work?

When your friend used both ears to hear, she likely guessed the correct direction of the sound more times than when she only used one ear to hear. There are two reasons for this. When your friend used both ears, the clicks reached the ear closest to the sound a split second before reaching the farther ear. In addition, the sound in the closer ear is slightly louder since the head is in the way of the farther ear. When one ear was covered, your friend couldn't use these slight differences in sound speed and volume to locate the sound of the pennies.

Humans have binaural hearing — that is, we use two ears to hear. A person with hearing in only one ear may have difficulty locating sounds. He might need to turn his head in several directions before finding the source of the sound.

TRY AN EAR TRICK

Trick a friend's ears with the penny clicker from the "Do We Really Need Both Ears?" experiment. Blindfold your friend. Click the pennies directly in front of her and ask her to locate the sound. Do the same directly behind her and on top of her head. Is your friend confused? She should be. In order to locate sound accurately, the sound should reach one ear sooner than the other. When the sound comes from a spot midway between your ears, it reaches both ears at the same time, making it tricky to figure out where the sound is coming from.

Q. Where should a farmer never tell secrets?

A. In a cornfield — it's all ears.

HEARING AIDS

AMPLIFY
SOUNDS

Some people need hearing aids to help them hear.
A hearing aid makes sounds louder
for a hearing-impaired person. Here's how it works.

You'll need:
- *a nail*
- *an empty shoe box*
- *2 paper fasteners*
- *scissors*
- *a rubber band*
- *a friend*

1 Use the nail to punch one hole at each end of the box. Put each hole midway between the sides of the box and 1 cm (1/2 inch) from the open top.

2 Insert the paper fasteners into the holes and open their stems on the inside of the box.

14

3 Cut the rubber band once to make a strip.

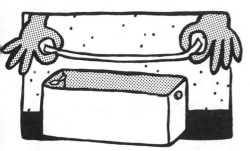

4 Holding the ends of the band, stretch the band so that it is about 4 cm (1 1/2 inches) longer than the length of the box.

5 Ask a friend to pluck the band with his finger. Note how loud it sounds.

6 Wrap one end of the band two or three times around one fastener underneath its head. Stretch the band across and wrap the other end around the other fastener.

7 Pluck the band with your finger. Is there a difference in the loudness of its sound?

? How does it work?

When only the rubber band is plucked, it vibrates and produces a soft sound. When the band is attached to the box and plucked, it makes the box and the air inside it vibrate, too. Since a much larger area is now vibrating, a much louder sound is produced. The box is acting as an amplifier, making sound louder.

A hearing aid also has an amplifier to make sounds louder. It has a tiny microphone that picks up sound and turns it into electricity. The sound, now in the form of electricity, is increased by an amplifier.

Then a tiny speaker, called a receiver, turns the electricity into sound that the person can hear.

Hearing aids sometimes work too well. In addition to picking up a person's voice, they also pick up all the other sounds in a room. If you wore a hearing aid in school, you would hear rustling paper, scraping chairs, and talking classmates almost as loudly as your teacher's voice. If you wanted to hear her voice more clearly, you and your teacher would each wear an FM (frequency modulator) on a string around your necks. The FM is like a one-way radio. It makes the teacher's voice stand out above the other sounds.

HEARING-EAR DOGS

If you were deaf, how would you know that the doorbell, your alarm clock, the stove timer, or the telephone were ringing? A hearing-ear dog could help you respond to these sounds. A hearing-ear dog is trained to nudge its owner whenever it hears one of these sounds. It then leads the owner to the source of the sound.

THAT DOESN'T SOUND LIKE ME

SOUND CONDUCTORS

Have you ever heard your voice
on a tape recorder and wondered if it were really you?
Try this experiment to find out why it doesn't sound like you.

1 Make a mallet by pushing the pointed end of the pencil firmly into the end of the cork.

2 Hold the fork between your thumb and two fingers near the top of the handle. Holding the mallet in your other hand, strike the tines of the fork as hard as you can. Listen to the sound it makes.

3 Strike the fork again. Then put the end of the fork handle between your front teeth. Be sure to hold on to the fork while you do this. Is there a difference in the loudness of the sound?

You'll need:
- *a pencil*
- *a bottle cork*
- *a metal fork*

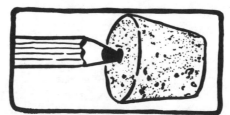

❓ How does it work?

When you hold the handle between your teeth, the sound of the vibrating fork travels through your jawbone to the cochlea (see diagram on page 11) embedded in your skull bone. Bones, or solids, are better conductors, or transmitters, of sound than air is. When you speak, you hear your voice through your skull and jawbones as well as through your ears. When you hear your voice on a tape recorder, you are listening to a sound that travels through air only.

People who are born without ears or an eardrum use a bone conduction hearing aid. The hearing aid is attached to a headband and lies against the mastoid bone behind the ear. Sound that is picked up by the hearing aid is amplified (made louder) and then travels through the skull bones to the inner ear.

What happens if...

- you press the vibrating fork handle against your mastoid bone, the flat bone behind your ear?

- you press the vibrating handle against your forehead? your cheekbone? your jawbone?

BEETHOVEN USED HIS TEETH TO COMPOSE MUSIC

Did you know that Beethoven, a famous composer of the late 1700s and early 1800s, continued to compose great music even after he became deaf? To help him hear the music he was writing, Beethoven would hold one end of a wooden stick between his teeth and put the other end against the piano strings. When he played a note, the sound travelled through the stick, through his teeth and skull bones directly to his inner ears.

DESIGN A BETTER HUMAN EAR

Think of the different ear shapes that you have seen —
cat ears, rabbit ears, elephant ears, human ears.
Which ear shape is the best for collecting sound? Could you
design a better human ear? Here's one idea to get you started.

You'll need:
- *a serrated knife*
- *an empty plastic cooking oil bottle*
- *an A tuning fork (440 cycles per second)*

1 With an adult's help, cut off a bit of the neck and the bottom of the bottle.

2 In a quiet room, strike the tuning fork on a soft, but solid, surface, such as the heel of your sneaker. Hold the vibrating fork about 28 cm (11 inches) from one ear and notice the loudness of its sound.

3 Put the neck end of the bottle to your ear. Strike the tuning fork again and hold it at the other end of the bottle at the same distance from your ear as in Step 2. How do the two sounds compare?

4 Repeat Step 3, but move the vibrating fork back and forth across the end of the bottle. How do humans catch sound that moves? How do animals do it?

? How does it work?

Sound travels through air in waves, which spread out in circles in all directions. When you listened to the fork without the bottle, some of the sound reached your ear, but most of it spread out into the room and was lost to your ear. When you held the fork near the end of the

bottle, the bottle channelled more sound towards your ear. The hum of the fork was louder since less sound spread out into the room.

Most animals have ears that are better designed to pick up sound than human ears are. Animals tend to have more cup-shaped outer ears to help them pick up sounds made by their enemies and prey. They also have the advantage of being able to turn their ears, which helps them locate the direction of sounds. Humans must turn their whole head to do this. Now, what do you think? Would you like to change the shape of your ears?

What happens if...

- you use a milk jug instead of the bottle?
- you use cones made out of bristol board? Try different depths and sizes.

OWLS HAVE LOPSIDED EARS

Imagine if you had one ear higher on your head than the other. Owls, sitting high up in trees, depend on this lopsided arrangement to find their food when hunting at night. Since one ear is lower than the other, sound coming from an animal below arrives at the lower ear a split second sooner than at the other ear. This helps the owl locate its dinner.

SHAKE 'N' TAKE

How sharp is your hearing?
Make and play a game of Shake 'n' Take to find out.
This game is for two to four players.

You'll need:
- *20 empty 175 g (6 ounce) yogurt containers with lids or 20 empty 280 mL (8.4 ounce) pop cans*
- *pencil*
- *paper*
- *scissors*
- *sticky tape*
- *a variety of materials to make shakers: e.g., rice, macaroni, popcorn seeds, bean seeds, paper clips, buttons, dried peas, barley, salt*

To set it up:

1 Using the bottom of a container, trace 20 circles onto the paper. Cut them out.

2 Line the circles up in pairs. Write the same letter of the alphabet on each pair, so that you have two A circles, two B circles and so on.

3 Tape the 20 circles to the bottoms of the containers with the letter side facing out.

4 Into each pair of containers, put the same amount of one kind of shaker material.

5 Put the lids on the yogurt containers or use paper and tape to cover the holes in the cans.

6 Shake each pair to make sure they sound the same. Also check that each pair sounds very different from the others.

To play:

1 Move the containers around so that the pairs are mixed up.

2 Set up the containers so that there are five rows with four cans in each row.

3 The first player shakes one container and then shakes another. If the sounds match, she keeps the pair. If they don't, she returns them to their original spots. If you're not sure that two sounds match, check the bottoms of the two containers.

4 The next player does the same. She may shake a container that the first player used or she may try two new ones.

5 The game continues until all the pairs have been found. The player with the most pairs wins. If you want to make the game even more challenging, use more shakers.

NOISE POLLUTION

We're all concerned about water, air and land pollution. But what about noise pollution? People living in large cities are surrounded by noise pollution. Every day their ears are bombarded by sounds made by vehicles, machines, television, speakers, planes and so on. As they grow older, many people lose some of their hearing. People who live in rural Africa have excellent hearing since there is very little noise to damage their ears.

Loudness of sound is measured in decibels. Listening to noise above 140 decibels (such as a jet taking off close by) over a period of time can cause permanent deafness. Here are some other ways in which our hearing can be damaged by loud sound.

● Some toy guns emit sounds of 170 decibels and some music played through speakers is as loud as 130 decibels.

● The noise level of a live rock concert is as loud as a jet plane taking off. Listening to a live rock band as often as once or twice a week can result in hearing loss. This means, among other things, that a person can't hear consonants such as f, s, k, p, h and g when someone is speaking or hear high notes in music.

● If you play your portable cassette player with the earphones on louder than half its possible volume on a regular basis, you are damaging your inner ear. Over time, this can result in hearing loss.

THE TRAVELLING SOUND SHOW

What can travel and bounce, yet can't be seen? The answer is sound.
Sound can travel through just about anything — air, wood, string, metal,
water and even bone. It can also bounce off hard surfaces to create echoes.
How does sound do all this? You'll find out when you travel through
the experiments in this section. You'll also discover how sound
can shatter glass, how it can be affected by instant coffee
and how it can help elephants to mate.

A TRAVELLING CONTEST

Sound can travel through solids such as
wood and metal as well as through air.
It travels through each of these materials at different speeds.
Which do you think is the best transmitter of sound —
air, wood or metal? If you're not sure, try this experiment and find out.

You'll need:
- *a wristwatch that ticks*
- *a metre (yard) stick*
- *a wooden table*
- *2 metal tubes from a vacuum cleaner*

1 In a quiet room, hold the watch away from one ear at a distance where you can't hear it. Move the watch towards your ear until you can just hear it. Stop. Measure the distance from your ear to the watch.

2 Lay the watch on a wooden table. Cover one ear. Press the other ear against the surface of the table about 60 cm (24 inches) away from the watch. Can you hear the ticking?

3 Move the watch further from your ear and listen again. What is the greatest distance over which you can hear the watch through the wood? Does sound travel better through air or wood?

from the sound source to your ear.

All things, including air, are made up of tiny particles called molecules. The molecules in solids, such as wood and metal, are more tightly packed than the molecules in air. Because the molecules in solids are so tightly packed together, they can carry sound waves more quickly and efficiently than the spread-out molecules of air. Also, air molecules are easily disturbed by other factors, such as wind or a person walking by, so they don't carry the sound waves as efficiently as solids do. That's why you heard the watch ticking from a greater distance through the wood and metal than through the air. Solids are good carriers, or transmitters, of sound. Sound travels through steel about 15 times faster than through air.

4 Connect the two metal tubes. Lay them on the floor. Cover one ear. Press the side of one end of the tube against your ear. Have a friend hold the watch by its strap against the side of the other end of the tube. What is the greatest distance over which you can hear the watch? You may have to add a third tube to find out.

❓ How does it work?

Sound travels in waves similar to the circles of water when you drop a pebble into a pond. The waves spread outward because the first wave pushes against the next one which pushes against the next one and so on. Sound travels like this

SNAKES HEAR WITHOUT EARS

Snakes use the fact that sound travels better through solids than through air to "hear" an approaching enemy or meal. Snakes have no ears, but if a snake lays its head on the ground, a bone inside its head picks up the sound vibrations coming from an approaching animal's movements. The vibrations travel to the snake's brain via a cochlea similar to the one inside the human ear.

AN UNUSUAL AMPLIFIER

SOUND TRAVEL

If you stick your fingers in your ears, do you think you'll be able to hear better? Impossible? Not if you get the sound to travel to your ears through string rather than through air.

You'll need:
- *a piece of string 90 cm (36 inches) long*
- *a wire clothes hanger*
- *2 metal spoons*
- *a metal fork*

1 Tie the middle of the string to the hook of the hanger.

2 Wind each end of the string twice around the end of each of your index fingers.

3 While you hold the hanger in mid-air, ask a friend to tap the hanger a few times with a spoon and notice the sound you hear.

26

7 Have her tap them again while you keep your fingers with the string in your ears. What is the difference in sound?

? How does it work?

When the hanger is tapped the first time, it vibrates and so makes a sound. The sound vibrations travel through the air in all directions. Since only a small part of these waves comes to your ear, the sound is not very loud. When your fingers are in your ears, the hanger's vibrations travel through the string directly to your ears. The sound is much fuller and louder. Sound travels much better through a solid such as string than through air.

5 Remove the hanger from the string. Tie the fork and a spoon to the string instead.

6 Have your friend tap the fork and spoon with the other spoon while you hold the string in mid-air.

4 Put your index fingers in your ears and lean forward so that the hanger hangs freely. Ask your friend to tap the hanger again. What is the difference in sound?

SOUND IN OUTER SPACE

What's the quietest place in the universe? Outer space. Sound must travel through molecules of air, liquid or solid in order to be heard. Since there are no air molecules in space, there is nothing to carry the sound waves. That's why astronauts can't talk to each other in the usual way; they must use radios to communicate. Radio waves *can* travel where there are no air molecules.

CHANGING PITCH

When a musician plays a melody on an instrument,
you hear notes that go up and down.
In other words, you hear a continual change in pitch.
But you don't need an instrument to change the pitch of a sound.
You can do it by changing the speed of the sound.
Try these two amazing kitchen experiments to find out how this is possible.

Stirring up sound in a cup of coffee

You'll need:
- *a mug*
- *boiling water*
- *2 metal teaspoons*
- *instant coffee*

1 With an adult's help, fill the mug three-quarters full with boiling water.

2 Put a spoon into the water and clink its edge against the side of the mug. Listen to the pitch (how high or low it sounds) of the clinking spoon.

3 As you continue clinking, pour one spoonful of instant coffee into the water. What happens to the pitch of the clinking?

4 Continue clinking for another minute. How does the pitch change?

? How does it work?

The particles of coffee powder have tiny air bubbles attached to them. When the coffee dissolves in the hot water, the air bubbles are released into the water. Since sound travels through air about four times more slowly than through water, the air bubbles slow the sound waves down. Slower sound waves give a lower pitch. That's why the pitch of the spoon clinking against the mug is lower when you first add the coffee. As you continue to clink and stir up the water, the coffee dissolves and the air bubbles rise to the top and escape. The pitch gradually rises, returning to the original pitch when the sound waves were travelling through water only.

What happens if...

● you tap the outside of a mug of water with a metal spoon while someone blows bubbles in the water with a straw? Listen to the pitch with and without the bubbles.

DOES SOUND TRAVEL THROUGH WATER?

The next time you go swimming, ask a friend to tap two spoons or rocks under the water while your head is above the water. Can you hear anything? Now put your head under the water and try it again.

Does sound travel through water? It sure does. In fact, it travels about four times faster through water than through air.

The wailing bowl

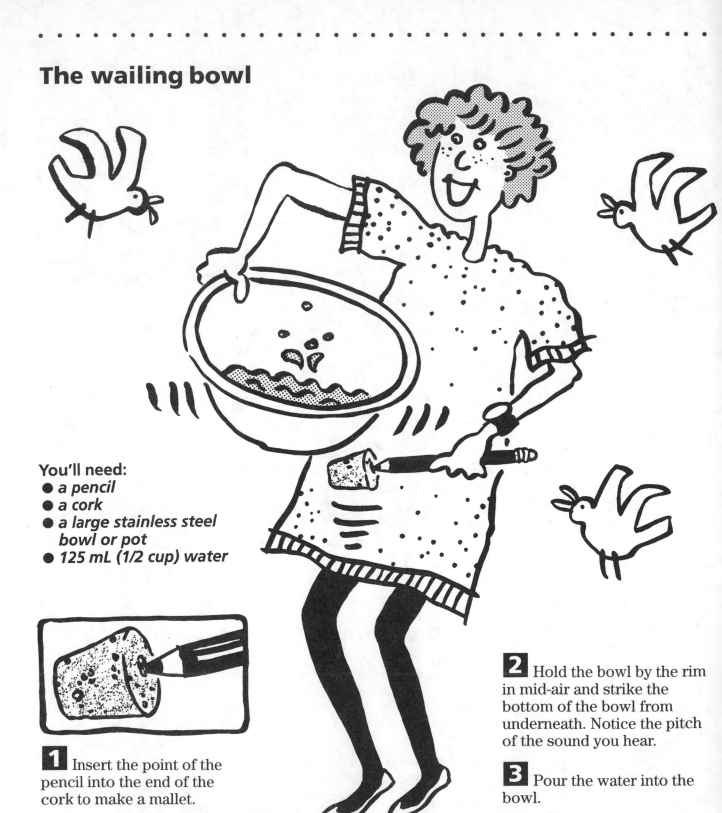

You'll need:
- a pencil
- a cork
- a large stainless steel bowl or pot
- 125 mL (1/2 cup) water

1 Insert the point of the pencil into the end of the cork to make a mallet.

2 Hold the bowl by the rim in mid-air and strike the bottom of the bowl from underneath. Notice the pitch of the sound you hear.

3 Pour the water into the bowl.

4 Hold the bowl steady in mid-air and strike the bottom with the mallet. Compare the pitch of this sound to that of the empty bowl.

5 Strike the bottom of the bowl again and tip it to one side. What change in pitch do you hear? Now strike the bowl and swish the water from side to side.

❓ How does it work?

When you strike the bottom of the bowl, it vibrates at a certain frequency or speed. The frequency with which it vibrates determines the pitch of the sound you hear. When you hit the bowl with the water in it, the pitch of the sound is lower because the water slows down the vibrations of the steel bottom. When you tip the bowl, the water leaves the bottom free to vibrate at a higher frequency and pitch similar to the pitch of the bowl without any water. The wailing sound you hear when you swish the bowl from side to side is a rising and falling of pitch when the water uncovers the metal bottom and covers it again and again and again.

SECRET LOVE CALL

Until recently, the love call of a female elephant was a secret that only elephants knew about. Even though bull (male) elephants can hear the love call of a cow (female) several kilometres (miles) away, the sound is too low for humans to hear. Using modern equipment, scientists have found a way to record the female's special call. Scientists hope that this information will help them to breed elephants in zoos. Sound that is so low that it can't be heard by the human ear is called infrasound.

Q. How can you tell that a brontosaurus is sleeping?

A. by the dino-snores

31

SEEING IS BELIEVING

Sound is something you hear but don't see. So, if you tell a friend that you can show her what sound looks like, she might not believe you until you show her this experiment.

You'll need:
- *glue*
- *a sheet of lined paper*
- *a piece of cardboard the same size as the paper*
- *a thick rubber band*
- *a felt-tip pen*
- *a wooden metre (yard) stick*
- *a friend*

1 Glue the sheet of paper onto the cardboard.

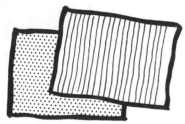

2 Use the rubber band to attach the pen to the metre stick so that the tip of the pen extends beyond the end of the stick by 2.5 cm (1 inch).

32

3 Lay the metre (yard) stick on a table so that the end with the pen extends beyond the edge of the table by 30 cm (12 inches).

4 Ask your friend to press down on the stick to keep it still on the table.

5 Holding the paper with the lines running vertically (up and down), touch the pen tip to the left side of the paper. Depress the metre stick as far as possible and let go. As the stick vibrates, move the paper quicky across the pen tip. You should have a series of waves.

6 Move the metre (yard) stick so that it extends 44 cm (18 inches) beyond the edge of the table. Do the same thing as in step 5. Count the number of waves (one wave is a down and up tracing) between four blue lines. How does the number of waves differ between steps 5 and 6?

7 Listen to the sound of the metre stick when it vibrates in each of the above positions. In which position is the sound lower? When does the stick vibrate more slowly?

? How does it work?

Sound can be represented on paper as a curved line or wave. A wavelength is the length of one complete wave. One wavelength is the distance between one crest, or high point, and the next one. The faster an object vibrates, the greater the number of wavelengths per second and the higher the pitch. In step 5, there were more wavelengths between the four blue lines than in step 6, so the pitch was higher.

A machine called an oscilloscope allows people to look at sound wave patterns on a screen. Many professionals use this machine to help them with their jobs. For example, police sometimes use an oscilloscope to identify people by their voice patterns.

A SILENT WHISTLE?

Can your dog add or subtract? Some performing dogs can do this trick when they are trained with a silent whistle that only dogs can hear.

When a trainer asks the dog, "What is 2 plus 2?", the dog barks four times because someone off-stage blows the silent whistle four times. Since the sound of the whistle is too high for the human ear to hear (people can't hear sounds higher than 20 000 hertz or 20 000 vibrations per second), the audience is impressed with the cleverness of the dog. Dogs have no trouble hearing the whistle since they can hear sounds as high as 50 000 hertz.

ECHOES

If you stand about 18 m (20 yards) away from a brick wall and shout towards it, you will hear an echo. An echo is caused by sound waves bouncing off the wall and returning to your ears. On a sunny day, try this experiment to find out what bouncing sound waves look like.

You'll need:
- *water*
- *a metal cookie sheet with edges*
- *a sharp pencil or a toothpick*
- *scissors*
- *a large round plastic container (such as an ice-cream tub)*

1 Pour water into the cookie sheet to a depth of 1 cm (1/2 inch).

2 Place the pan in a sunny spot near a window so that the sunlight is reflected by the water onto the ceiling. You should see the shape of your pan shining on the ceiling.

3 Wait until the water in the pan is completely still.

4 Touch the point of the pencil to the surface of the water in the middle of the pan. What do you see in the ceiling reflection?

34

5 What do you see when you touch the water in different parts of the pan?

6 Cut the rim off the plastic container. Bend the rim so that it is in the shape of a slight curve.

7 Place the rim in the water at one end of the pan. With the rim in the pan, touch the pencil to the surface of the water again. What differences do you see in the movement of the waves?

❓ How does it work?

When you touch the water with the pencil, the water waves go out in circles. When they hit the sides of the pan, they bounce off the sides and move back towards the middle. This is exactly what happens in the air when you hear an echo. The sound waves coming from your shout hit a hard surface, such as a brick wall, and bounce off it. When the sound waves come back to you, you hear an echo of your shout.

When the water waves bounce off the curved plastic rim, they meet at a certain spot in the pan rather than travel straight across to the other side. This happens with sound in rooms called whispering galleries — rooms with domed ceilings and curved walls. If you stand near the wall of a whispering gallery and whisper towards it, the domed ceiling and curved walls of the room reflect your whisper in such a way that a person far across the room can hear you clearly. St. Paul's Cathedral in London, England, and the Mormon Tabernacle in Salt Lake City, U.S.A., are known for their whispering galleries.

SINGING IN THE SHOWER

Do you sound best when you sing in the shower? Why is that? When you sing in the shower, sound waves travel from your throat towards the wall of the shower stall. They bounce off the hard tiles and travel to the opposite wall, which also reflects the sound waves. This bouncing back and forth happens continually while you are singing, making your voice sound fuller and richer than it really is.

RESOUNDING SOUND

DISCOVER RESONANCE

You can play the piano without touching a single key. Sound impossible? Not if you use some resonance.

You'll need:
- *water*
- *an empty 750 mL (24 oz.) pop bottle*
- *a friend*
- *a piano*

1 Pour water into the bottle until it is half full.

2 Practise blowing across the mouth of the bottle until you get a clear tone. To learn how to do this, see Pop Music on page 65.

3 Ask your friend to open the piano lid and press down the pedal on the right. This lifts the dampers off the strings so that they can vibrate freely.

4 Stand beside the piano and blow across the top of the bottle for as long as you can. Then listen to the sound coming from the piano strings. Can you guess which strings are vibrating? Touch the strings gently to find out.

5 Pour more water into the bottle until it is two-thirds full and blow again. What is the difference in sound? Try blowing softer and louder tones. What do you hear from the piano?

How does it work?

When you blow across the top of the bottle, the air inside the bottle vibrates at a certain frequency or times per second. This vibrating air travels to your ear as sound. It also travels to one set of piano strings and makes it vibrate. Why that set and not another? That set of strings has the same pitch, or vibrates at the same speed, as the air in the bottle. Scientists say that the bottle sound is in resonance with the particular piano sound. A different sound from the bottle will be in resonance with another set of piano strings.

Interesting things can happen when two things are in resonance. Every object has a rate at which it vibrates naturally. An opera singer can shatter a glass when the note she sings is in resonance with the natural vibrations of the glass.

What happens if...

● you use your voice instead of the bottle?

● several people sing in harmony?

● you use an instrument such as a xylophone or recorder instead of the bottle?

WHAT'S IN A SEASHELL?

When you hold a shell to your ear, you might think you hear the ocean. What you really hear is the air inside the shell vibrating in resonance with the sounds around you. What happens when you hold different sized cans to your ear instead of a seashell? Try it in a variety of locations in your home and neighbourhood.

Q. Why did the boy take a ladder to music class?

A. His teacher wanted him to sing higher.

SOUND CHARADES

SOUNDS LIKE FUN!

The next time you have some friends over, challenge them to a game of Sound Charades.

A Horse!

A galloping horse!

You got it!

You'll need:
- *30 or more blank cards made of cardboard or paper*
- *a pencil*
- *objects that can be used to make sound effects (see page 72 for ideas)*
- *a watch*
- *paper for score keeping*

1 On each card write the name of one object, animal or action that makes a sound. A few ideas are listed on page 39. Shuffle the cards and lay the deck face down.

2 Divide the players into two teams. Each team sits on either side of the deck of cards.

3 The first person on Team One picks the top card. His teammates turn their backs to him so that they can't see him.

4 The first person has one minute to get his teammates to guess the sound on the card. He may not talk, but he may use his mouth, body or any helpful objects to make appropriate sounds.

5 If someone on Team One guesses correctly before the minute is up, that team gets one point.

6 Team Two uses the watch to keep time. When the time is up, it is Team Two's turn to pick the next card and play.

7 The game continues until one team reaches a score of ten. Here are some sound ideas for your cards:

–a doorbell	–a motorcycle
–a violin	–a rainstorm
–a horse	–a train
–a bass fiddle	–a heartbeat
–ocean waves	–fire
–Morse code	–a clock
–a trumpet	–a seal
–a mosquito	
–water from a tap	
–popcorn popping	
–a squeaky see-saw	

EAR, THERE AND EVERYWHERE

In the animal world, you can find creatures that have hearing organs on every imaginable part of their body. Crickets have membranes like eardrums on their thighs. Most spiders pick up sound vibrations with the hairs on their legs. Tarantulas feel vibrations with the soles of their feet. Fish have two sets of hearing organs — they have hearing sacs in the head as well as a tube that runs along each side of the body. Lining these sacs and tubes are tiny hairs that pick up sound vibrations from the water. Imagine having ears in these places on *your* body!

WHISTLE WHILE YOU MOW

If you're not an expert whistler, use some grass to help you. Find a long, wide blade of grass. Put the grass blade between your thumbs so that it is taut. Hold your thumbs side by side with the nails facing you. Blow through the small space between your thumbs.

PUT SOUND TO WORK

What do a farmer, a doctor, a police officer and a ventriloquist
have in common? They all use sound to help them do their jobs.
In this section, you'll find out what kind of music plants like best,
how stethoscopes work and how police officers catch speeders with sound.
You'll also get a chance to experiment with ventriloquism
and try out some sound techniques that blind people use to get around
in their daily lives. So read on and put sound to work.

GROW PLANTS WITH SOUND

What kind of music do you like?
Some botanists (plant scientists) claim that certain types of music and sounds can stimulate plants to grow better.
Try this experiment to find out what kind of music your plants like.

You'll need:
*4 identical flower pots with plates or pans to set them on
potting soil
a trowel
lima bean seeds
water*
● *paper and pencil
plant fertilizer that contains trace elements such as zinc, magnesium, iron
a spray bottle (mister)
a tape recorder or record player*

1 Fill each pot with an equal amount of potting soil. For a good scientific experiment, you must treat each plant the same way.

2 Plant two seeds that have been soaked in water overnight in each pot. Plant the seeds about 2.5 cm (1 inch) deep.

3 Water each pot until a few drops drain out of the holes at the bottom.

4 Label the pots:
1 - violin music   3 - bird song
2 - rock music   4 - control

7 Do the following every day either before 11 a.m. or after 4 p.m. Put pots 2, 3 and 4 in another part of your home. Spray the leaves of pot 1 with the fertilizer. This is called foliar (leaf) feeding. Next to pot 1, play recorded violin music for half an hour. Use music that is dominated by violins since it is the high pitch of violins that will affect your plants.

8 Now replace pot 1 with pot 2 and play rock music for pot 2. Do the same for pots 3 and 4, playing recorded bird song for pot 3 and no music for pot 4. Be sure to spray each plant with the same amount of fertilizer and keep the volume and the distance of sound from the plants the same for each pot. Return all the pots to the window after their music session.

9 Keep a record of the growth of your plants. You might want to record their height, the number of leaves, the size of leaves, the shade of green, the thickness of the stems.

10 At the end of your experiment (as many weeks as you wish), decide which sound is most effective for growing your plants.

5 Place the pots in a sunny window. Water them whenever the soil is dry until drops come out the drainholes. Remember to use the same amount of water for each plant.

6 As soon as two leaves appear on the plants (in about two weeks), start your experiment. Pour a diluted solution of fertilizer into the mister. (To dilute the fertilizer use half the suggested amount of fertilizer to the required amount of water.)

❓ How does it work? 🎵

The leaves of plants are covered with tiny openings called stomata. While stomata normally release oxygen and absorb carbon dioxide, they can also absorb nutrients when sprayed with diluted fertilizer. Certain sounds stimulate the stomata to open and absorb the nutrients that are used by the plant to help it grow. Some botanists and farmers have found that bird song, violin music, Hindu ragas and whistling are most effective in improving plant growth. Corn and wheat farmers in the United States have reported greater yields per hectare (acre) when their crops were stimulated by Bach violin sonatas.

LISTEN TO THE BEAT

MAKE A STETHOSCOPE

When a doctor listens to your heartbeat with a stethoscope, she is actually listening for two sounds. The first sound is a low-pitched longer sound — lubb. The second is a higher, more snapping sound — dup. Find out what these sounds mean by making and using this stethoscope.

You'll need:
- *a 30 cm (12 inch) length of aquarium tubing that fits snugly over the spouts of the funnels*
- *2 small funnels (the smaller the better)*
- *a balloon*
- *scissors*

1 Fit the ends of the tubing over the spouts of the funnels.

2 Inflate the balloon and hold it closed for two minutes to stretch it. Let out the air.

3 Cut off the top third of the balloon so that you have a small cap.

4 Stretch the cap as taut as possible over the open end of a funnel.

5 In a quiet room, place the capped funnel flat against your chest to the left of centre. (Your heart sounds loudest here because its bottom tip touches the chest wall at this point.) Hold the other funnel over your ear.

6 You will hear a low lubb-dup. Count the number of lubb-dups you hear in 15 seconds. Multiply this number by 4 to find out how fast your heart beats in one minute.

7 Jump up and down for three minutes. How fast does your heart beat now?

8 Listen to a baby's heart, your parent's heart, and a pet's heart when they are at rest. How do their heart rates compare to yours? How do their heart rates change after exercise?

? How does it work?

The low-pitched longer dubb sound is made by the closing of two heart valves when blood is flowing out of the heart. The more snapping dup sound is made by two other valves when blood is flowing into the heart. When a person exercises, the heart beats faster in order to pump more blood and oxygen to the muscles that are being used.

The sound of the closing of the heart valves causes the stretched balloon to vibrate. The vibrating balloon makes the air in the tube vibrate and the tube carries these sound vibrations to your ear.

A real stethoscope uses a thin plastic disc or diaphragm instead of a balloon in the chestpiece.

Two tubes extend from a shorter single tube to reach both ears. A doctor also uses a stethoscope to listen to your lungs, stomach and intestines.

What happens if...

- you listen to a heart with a paper towel roll?
- you use the stethoscope without the balloon?

THE FIRST STETHOSCOPE

Before 1816, a doctor would listen to a person's heart by putting his ear to the patient's chest. In 1816, a French doctor, René Laennec, had trouble hearing a girl's heart with his ear because she was extremely overweight. He rolled up a sheaf of papers and put one end to her chest and the other end to his ear. The heart sounds were loud and clear. He improved this invention by making a wooden tube with one end in the shape of a funnel. This was the first stethoscope. It wasn't till 1855 that it started to look like the modern stethoscope with flexible tubing for both ears.

SEEING WITH SOUND

Which of your five senses do you think
you use most in your daily activities?
If you guessed your sense of sight, you were right.
When a person loses her sight, she must learn to use her hearing to do
the many things for which she once used her eyes.
There are three main listening skills that a blind person learns to use:
localizing sound, discriminating sound and echolocation.
Here's your chance to try out a few of these listening skills.

You'll need:
- *a blindfold*
- *a friend*
- *some keys on a ring*
- *a radio*

Localizing sound

1 In a quiet room, sit in front of a table and blindfold yourself. Make sure your ears are uncovered.

2 Have a friend drop the keys somewhere on the table. Use your hearing to find them.

3 Do this several times, dropping the keys in different places on the table.

4 Now sit in the middle of the room. Have your friend jingle the keys somewhere in the room. Can you point in the direction of the sound? Do this several times in different parts of the room. Do you get better with practice?

● A blind person must be able to accurately locate sounds. This is especially important on a busy street corner. She needs to know where the cars are moving and where they are stopped in order to cross the street safely.

Discriminating sound

1 Turn your radio to a program in which people are talking.

2 While you are blindfolded and with the radio playing, have a friend tell you a story.

3 Turn off the radio. Repeat the story to your friend. Include all details.

● Was it hard to pay attention to your friend's story while the radio was on? A blind person must concentrate on sounds important to her while filtering out other sounds. If she is waiting at a bus stop, she must be able to hear an approaching bus and ignore other traffic sounds.

Echolocation

1 While you are blindfolded, have your friend lead you into different sized rooms in your home. You can even try a walk-in closet. Stand in the middle of the room and clap your hands twice. Can you tell from the sound of your claps whether you are in a small, medium, or large sized room? If not, try it again.

2 While blindfolded, stand about three metres (yards) in front of a large brick wall outdoors. The side of a garage or your school will do.

3 Clap your hands as you walk towards the wall. Listen carefully for differences in sound as you approach the wall. Can you use those differences to stop yourself from bumping into the wall?

● The sound from your clapping hands bounces off or echoes from the walls around or in front of you. The closer you are to the walls, the faster the echo returns to your ear. This explains the differences in sound that you hear.

Instead of hand claps, a blind person listens to the taps of her cane to tell her how far she is from an object. This is called echolocation. If you had difficulty hearing the differences in echoes, don't worry. It takes a lot of time and practice to use echolocation accurately.

AMAZING BATS

Bats can find an insect as thin as a hair in total darkness! How do they do it? They use echolocation. Bats make squeaking sounds that bounce off insects near them. The echo from an insect tells a bat how far away the insect is, as well as its shape and size. Using echolocation, a bat can catch as many as 600 mosquitoes in only one hour.

THE DOPPLER EFFECT

Have you noticed how the sound of a speeding car or train changes when it passes by you? In the 1800s, Christian Doppler — an Austrian mathematician — discovered why a sound seems to change from a high pitch to a low pitch when it passes a person who is standing still. Here's how it works.

You'll need:
- *2 m (6 feet) of strong string*
- *an alarm clock with a bell or a watch with an alarm*
- *a friend*

1 Do this experiment in an open space outdoors. Tie one end of the string very securely to the clock.

2 Set the alarm to go off. Have your friend stand at a safe distance away from you (more than 2 m [6 feet] away).

3 When the alarm starts ringing, whirl the clock in circles over your head. As you continue to whirl, let more of the string out until you are holding it near its end. What do you hear? What does your friend hear?

4 Now switch places with your friend and repeat the experiment. Does the alarm sound different to you than it did when you were swinging the clock?

? How does it work?

The sound of the alarm travels through the air in sound waves that spread out in all directions. When you whirl the clock above your head, the alarm always sounds the same because it is always an equal distance from your ears. Its sound waves reach your ears evenly. When your friend whirls the clock and you stand back, the distance between your ear and the clock is constantly changing. When the clock approaches you, the sound waves are bunched together so more waves hit your ear per second. That's why the alarm sounds higher. When the clock swings away from you, the sound waves are stretched out and hit your ear less frequently. The pitch of the alarm sounds lower. This change in sound or wave frequency is known as the Doppler effect.

SPEED TRAPS

Doppler's discovery is used today by police to catch speeding drivers. A radar gun, attached to a stationary police car, sends out radio waves. When the radio waves hit a moving car, they bounce off the car back to the radar gun just like an echo bounces off a wall. The faster the car is moving, the more frequently the radio waves hit the radar gun. A machine attached to the gun translates the frequency of these sound waves into a speed written in digital numbers.

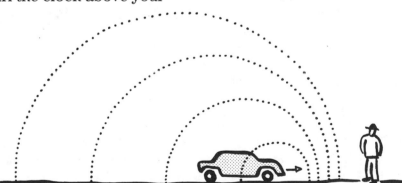

When a moving car approaches you, its sound waves are bunched up in front, and they reach your ear frequently, creating a high-pitched sound. As the car passes and speeds away from you, the sound waves behind the car are more stretched out. The waves reach your ear less frequently and the sound is lower pitched.

FIND THE SWEET SPOTS

Can you find the sweet spot on a baseball bat?
That's the place on the bat that will
most likely guarantee a homer if the ball makes contact with it.
Here's how to find it.

You'll need:
- *2 chairs with arm rests or leg bars*
- *string*
- *a piece of wood 1.5 cm (3/4 inch) thick, about 5 cm (2 inches) wide and 30 or 60 cm (12 or 24 inches) long*
- *salt*
- *a pencil*
- *a cork*

1 Stand the chairs so that they are about 30 cm (12 inches) apart.

2 Stretch a piece of string from one chair across to the other. Tie it to both chairs.

3 Do the same with a second piece of string so that it is parallel to and about 18 cm (7 inches) apart from the first.

4 Lay the wood across both strings so that it is evenly balanced.

5 Sprinkle salt all over the wood.

6 Push the point of the pencil into the end of the cork to make a beater.

7 With the beater, tap the wood continuously to keep it vibrating. Keep tapping until the salt forms a definite pattern. The places where the salt gathers are the sweet spots, or nodes.

50

? How does it work?

Since the wood is suspended on the string, it vibrates freely when you hit it with the beater. The jumping salt particles show that the wood is vibrating. The two spots where the salt gathers and stays are the spots, called nodes, where the wood does not vibrate.

A baseball bat has two nodes where it doesn't vibrate just like the piece of wood. The sweet spot on the bat is the node that is about three-quarters of the way up the bat. When you hit the ball at this spot, no energy is used for vibrating. All the energy goes into hitting that ball out of sight. That loud crack you hear when someone hits a homer is like the sound you hear when you strike the wood at a node. It sounds different than the other parts of the wood.

Xylophones work in the same way. Each bar on a xylophone is nailed down at its nodes. This allows the rest of the bar to vibrate freely and give that full rich sound.

DRAWING WITH VIBES

About 200 years ago, Ernst Chladni, a German scientist, discovered that he could "draw" patterns with sound. He sprinkled fine sand on a thin steel disc and stroked the edge of the disc with a violin bow. The continuous playing of a sound at a constant pitch caused the sand to form line patterns such as spirals, stars, honeycombs and concentric circles. Each time Chladni changed the frequency, or speed, of the violin bow, the pitch changed and different patterns formed. Chladni discovered that lines of sand formed in spots called nodes where the disc was not vibrating. From his experimenting, we know that all objects have nodes or spots that don't vibrate when the object is struck. The location of the nodes depends on the shape and material of the object as well as the frequency with which it is vibrating. Some violin makers find that violins that produce particularly good sounds have specific nodal patterns.

SOUND ILLUSIONS

SOUNDS LIKE FUN!

Ventriloquists use a combination of optical and sound illusions to make us think that their dummy is doing all the talking. With a little practice, you can do it too.

What animal has the best sense of humour?

I know, a laughing hyena! Ho! Ha! Ha!

1 When you make the puppet talk, hold your teeth about 1/2 cm (1/4 inch) apart and your lips slightly apart.

2 In front of the mirror, practise saying these letters one at a time without moving your lips: a c d e g h i j k l n o q r s t u y z. These are the easy sounds.

You'll need:
● *a puppet* ● *a mirror*

3 Since you must move your lips for the remaining letters, you need to substitute sounds or letters for them:

For b sound, say d or t instead.

For f sound, say th or h.

For m sound, make a slight hum followed by ng.

For p sound, say t or k.

For v sound, say thee

For w sound, say oooo.

Whenever you say a word with one of these letters at the beginning, say the latter part of the word louder and with more emphasis. For example, if you want your puppet to say "peanut butter," you must say "tea*nut* dut*ter*."

4 Make your puppet's voice higher or lower than your usual voice. After you've practised phrases and short sentences, make up a routine that you can perform for your friends. You might ask your puppet a series of questions about something special that he has done. He can answer you with smart-alec answers that will make your audience laugh.

? How does it work?

When a ventriloquist performs, he is sitting very close to his dummy. Since the human ear cannot pinpoint the exact source of the sound and since the dummy's lips are the only ones that are moving, it seems like the dummy is doing all the talking.

The other sound illusion that a ventriloquist uses involves the substitution of sounds. When you substitute d for b and t for p in peanut butter, the latter half of the words is emphasized to draw attention away from the sound of the first letter. If you said tea*nut* dut*ter* by itself, it might sound funny. But when it is in the context of a sentence, it sounds normal.

DO YOU HAVE BORBORYGMI?

From time to time, you probably have borborygmi. Don't worry. It just means that your stomach is growling.

Your stomach and intestines have very strong muscles that mash and grind the food that you eat. When you are hungry, there isn't much food in your stomach, only a bit of liquid and a lot of air. As the stomach starts the mashing process, the liquid and air get moved around and that's the growling sound you hear — and everyone else hears, too.

MUSIC TO MY EARS

You've probably heard of famous musicians who play either stringed, wind or percussion instruments. But what about idiophone or pop bottle players? Never heard of them? Well that's because there aren't many around. So here's your chance to shine. Once you've mastered some idiophones and pop bottles, you can figure out how stringed and wind instruments vary their pitch, why male singers sound different from female singers, and how air helps to amplify sound in various instruments.

A BOX FULL OF SOUND

What is empty and yet full of sound?
A sound box. Stringed instruments have
a sound box to make the sound of their strings louder.
To find out how this works, make and play this box harp.

You'll need:
- a pencil
- an empty tissue box (200-tissue size)
- 10 paper fasteners
- scissors
- 5 rubber bands, all the same size
- some cardboard
- tape
- glue

1 Use the pencil to punch five holes at each end of the box. The holes should be 1 cm (1/4 inch) from the top of the box. Measure 3.5 cm (1 1/2 inches) in from the side of the box and make the first hole. Make the other holes 1.5 cm (1/2 inch) apart in a line across the box.

2 Insert the fasteners into the holes and open their stems.

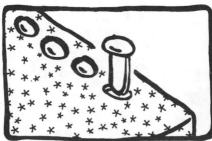

3 Cut each rubber band once to make a strip.

4 Tie one end of each strip around a fastener just below its head. Use a double knot.

5 Loop the other end of each strip twice around the fasteners at the other end. Do not tie them.

6 Cut two cardboard rectangles 5 cm by 15 cm (2 by 6 inches). Score each rectangle by drawing a line with the tip of a scissors along its length to make three equal columns.

7 Fold along the scored lines and tape to make two long triangular bridges.

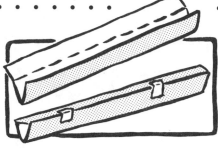

8 Slide the bridges under and across the rubber strings. Glue each bridge in place 2 cm (3/4 inch) from each end of the box.

9 Tune your box harp by stretching or loosening each string (rubber band) and rewrapping it twice around its fastener. Can you tune the strings so that they go from a low to high pitch?

10 Now you are ready to play your box harp by plucking the strings.

? How does it work?

As you pluck a string with one hand, gently touch the top edge of the box with the finger pads of the other hand. Do you feel the box vibrating? The vibrations of the string make the box and the air inside it vibrate. Since the sound box has a much larger surface area than the string, it can shake many more molecules of air. When more molecules are vibrating, you hear a louder sound. Guitars, violins, pianos and other stringed instruments use sound boxes to make their strings sound louder. Most sound boxes have a sound hole which allows the vibrating air inside the box to travel to the outside air and on to your ear.

What happens if...

- you use a larger box?
- you change the shape or size of the sound hole?
- you use a box made of thicker cardboard or another material?

Q. What kind of musician never gets locked out of his house?

A. A pianist. He always has a set of keys.

FEMALE VOICES — MALE VOICES

EXPERIMENT WITH VOCAL CORDS

Most female singers have higher voices than male singers. Why? You don't have to look down anyone's throat to find out. Just try this.

You'll need:
- *a medium-sized rubber band*
- *a medium-sized book*
- *2 pencils*

1 Put the rubber band lengthwise around the book so that there are no twists in the band.

2 Slide the pencils under the band so that each pencil lies across the book at either end.

3 Gently pluck the band with a finger. Listen to its sound. Can you see its vibrations?

4 Move one pencil to the middle of the book. Pluck the band between the two pencils. How is the sound different this time? How are the vibrations different?

58

? How does it work?

When you pluck the whole band, the vibrations are so slow that you can easily see them. Slower vibrations give a lower sound or pitch. When you pluck a shorter length of the band, the vibrations are so fast that you can hardly see them. Faster vibrations give a higher pitch.

One reason that female singers have higher voices is that their vocal cords are shorter than men's. The vocal cords are like two rubber bands that vibrate in the voice box when a person speaks or sings. You can feel these vibrations if you put your fingers on the bony part of your throat while you sing. You are touching your Adam's apple. The front ends of your vocal cords are attached to it. When you sing, the vibrations of the vocal cords are transferred to your Adam's apple.

The pitch of a sound is measured in hertz (Hz). Female singers with shorter vocal cords can sing as high as 1046 Hz. Male voices sing no higher than 466 Hz.

WHERE'S THE HUM IN A HUMMINGBIRD?

Why does a hummingbird hum? Because it doesn't know the words. Actually, hummingbirds hum with their wings, not their voices. They beat their wings backwards and forwards in a figure eight shape from 50 to 80 times in just one second. This rapid wingbeat makes the humming sound we hear. It also allows the hummingbird to hover in mid-air above a flower and to rise up like a helicopter when leaving a branch.

AN UNUSUAL TEMPERATURE GAUGE

Did you know that you can use your ears to tell the temperature of the air? On a warm summer night, count the number of cricket chirps you hear in 15 seconds. Divide that number by 2 and add 6. Your answer tells you the temperature in Celsius. To get a reading in Fahrenheit, simply add 40 to the number of chirps in 15 seconds. You can prove you're right by checking a thermometer.

This works only with tree crickets who live near and in wooded areas. Male tree crickets chirp by rubbing one hard wing against the other wing, which has tiny bristles along it. The males chirp to attract female crickets or to warn other males to stay away.

SWING WITH A STRING

Do you think you could play a one-stringed instrument?
How would you vary the pitch?
Make this unusual version of a bass fiddle and find out.

You'll need:
- a can opener
- an empty 1.36 L (48 ounce) juice can
- a nail
- a hammer
- heavy string
- a pencil

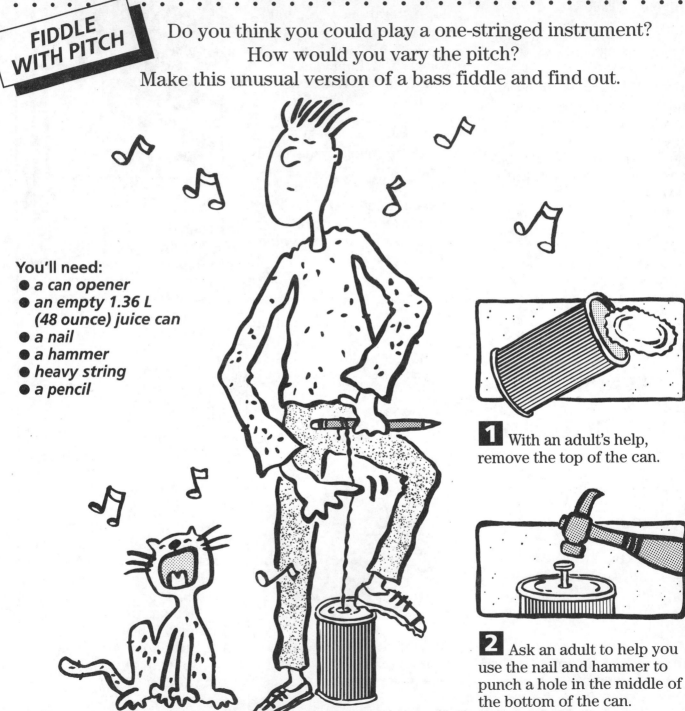

1 With an adult's help, remove the top of the can.

2 Ask an adult to help you use the nail and hammer to punch a hole in the middle of the bottom of the can.

3 Cut a length of string that stretches from the floor to the middle of your thigh.

4 Make a large knot at one end of the string. Pull the string through the hole in the can so that the knot catches on the inside.

5 Tie the other end of the string tightly to the middle of the pencil.

6 To play your bass fiddle, place the can on the floor in front of you and put one foot on top of it to hold it in place. Holding the pencil perpendicular to the string, curl your fingers over the pencil and use it to pull on the string. Pluck the string with the forefinger of your other hand. Change the pitch by loosening and tightening the string.

? How does it work?

When you pluck the string, it vibrates. When the string is tighter, it vibrates more quickly (at a higher frequency), and the pitch is higher. Since the looser string vibrates more slowly (you can see the vibrations), the pitch is lower.

Musicians tune stringed instruments such as guitars, violins and bass fiddles by tightening and loosening each string. Each string is attached to a tuning peg at the end of the instrument's neck. As the peg is turned, the string winds up or unwinds a little to get tighter or looser.

A ONE-MAN BAND

Can you beat the Guinness record for a one-man band? Eugenio Soler of Barcelona, Spain, developed a tricycle with moving parts on which he could play 50 instruments simultaneously. It was first demonstrated on June 21, 1988 (International Day of Music).

Q. What stringed instrument never tells the truth?

A. A lyre.

THROUGH THICK AND THIN

FREQUENCY AND PITCH

If you look at the strings on a violin or guitar, you'll find that they vary in thickness. In the last two experiments, you saw how the length and tension of the strings determine pitch. Try this experiment to discover how the thickness of strings affects sound.

You'll need:
- *a nail*
- *a hammer*
- *2 empty 540 mL (19 ounce) tin cans*
- *a thick piece of string 43 cm (18 inches) long*
- *a thin piece of string 43 cm (18 inches) long*
- *a piece of paper towel*

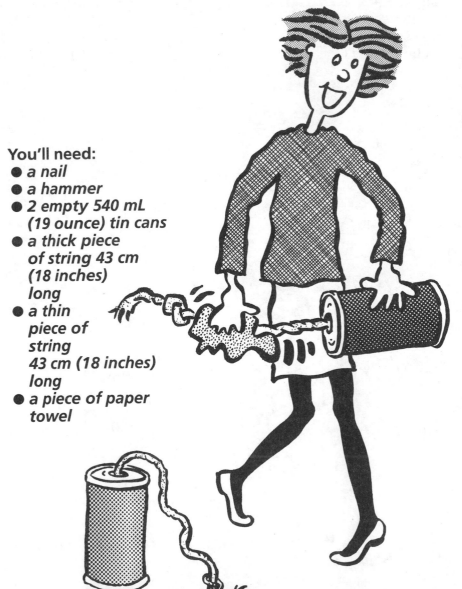

1 Ask an adult to help you use the nail and hammer to make a hole in the middle of the bottom of each can.

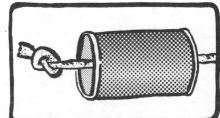

2 Make a large knot at one end of the thick string. Pull the string through the hole of a can so that the knot catches on the inside.

3 Do the same with the thin string and the other can.

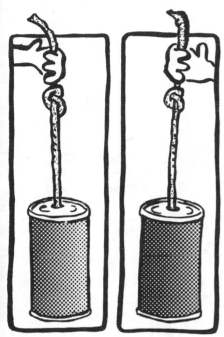

4 Make a single knot in each string 18 cm (7 inches) from the bottom of the can.

5 Moisten the piece of paper towel. Holding a can in one hand, press the string between the damp paper towel with your fingers. Starting at the bottom of the can, pull the paper along the string to the knot.

6 Do the same thing with the other string and can. Do you hear a difference in pitch (high or low sounds)? Why are the sounds of the strings so loud?

? How does it work?

As you pull your fingers along the string, it vibrates. The pitch of the thin string is higher because it vibrates more quickly (at a higher frequency). The can acts as a sound box to make the sound of the string louder than it would be if you pulled the paper along the string alone.

If you pluck each string on any stringed instrument separately, you'll find that the thinner strings produce higher notes than the thicker ones.

What happens if...

● you use other thicknesses of string?

● you use other sizes of cans?

● you use containers made of other materials?

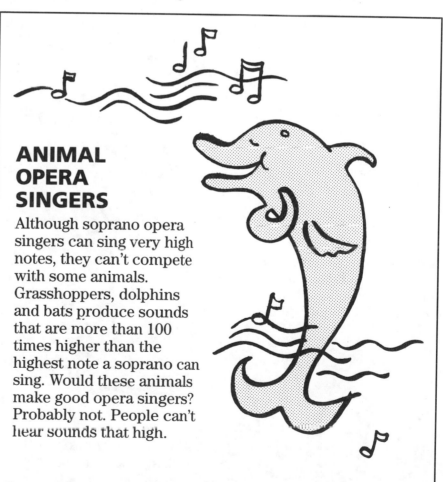

ANIMAL OPERA SINGERS

Although soprano opera singers can sing very high notes, they can't compete with some animals. Grasshoppers, dolphins and bats produce sounds that are more than 100 times higher than the highest note a soprano can sing. Would these animals make good opera singers? Probably not. People can't hear sounds that high.

POP MUSIC

Before you recycle those empty pop bottles, use them to create a new kind of wind instrument. When your family hears you, they might decide to keep those bottles forever.

You'll need:
- *5 empty pop bottles the same shape and size (750 mL [24 ounce])*
- *paper, pencil and tape*
- *water*

1 Line up the bottles near the edge of a table. Starting on the left, use small squares of paper to label each bottle: G A B C D.

2 Pour water into the G bottle so that it is about 12 cm (5 inches) high.

3 Pour water into the other bottles so that each bottle has about 2 cm (3/4 inch) more water than the one to the left of it.

4 Starting with the G bottle, blow across the top of each bottle in order. To get a good tone, raise your chin and put the edge of the rim at the bottom of your lower lip. Tighten your lips against your teeth and blow across the opening. Your breath should hit the opposite rim.

5 The notes you play should go up step by step. If they don't, adjust the water levels slightly. If you have a piano, you can use it to help you tune your bottles.

6 For your first musical performance, try "Jingle Bells": BBB BBB BDG AB CCC CC BB BBDDCAG.

7 Now try other songs. If you need a greater range of notes, add more bottles with different water levels.

? How does it work?

Sounds are made when something is vibrating. When you blow across the top of a bottle, the air inside the bottle vibrates. The more air in the bottle, the slower it vibrates and the lower the sound. That's why the bottle with the least amount of water (and the most air) has the lowest sound.

Flutes, oboes, clarinets and other wind instruments use a vibrating column of air to make their music. The higher notes in an orchestral piece of music are played by flutes, which have shorter air columns. The lowest notes are played by a contrabassoon, which is about 5 m (16 feet) long. To make it easier to hold and play, all that length is folded into three sections.

What happens if...

● you use smaller bottles?
● you blow across the top of a pen cap?
● you tap each bottle with a pencil instead of blowing? In this case, it's the glass that is vibrating.

THE NOISIEST ANIMAL IN THE WORLD

If you travel to the tropical forests of South America, you'll find an animal that makes the loudest sound that an animal is capable of producing. The male howler monkey has two bony "sound boxes" in its throat. When it howls, air blows across the open tops of the boxes just like air blowing across empty pop bottles. The resulting roar, which warns other howler groups to stay out of its territory, can be heard over a distance of 5 km (3 miles).

PLAY AN IDIOPHONE

Does your teacher's chalk scraping on the chalkboard sometimes send shivers up your spine? When it does, your teacher is playing an idiophone. "Idio" means self and "phone" means sound. An idiophone is anything that naturally makes sounds when it is rubbed, hit, shaken or scraped. Here's one idiophone that sounds more musical than your teacher's chalk.

You'll need:
- *water*
- *3 clean wine glasses*

1 Wash your hands thoroughly to make them free of grease.

2 Pour water into the glasses so that one is three-quarters full, the second is half full and the third is one-third full.

3 Wet a fingertip and rub it gently back and forth on the rim of each glass until it rings. What are the differences in pitch among the glasses?

4 Add some more glasses with other water levels and play a tune that you know.

? How does it work?

When you rub your finger on the rim, the glass vibrates. The water inside the glass slows the vibrations down. (If you look closely, you can see the water vibrating.) Slower vibrations give lower sounds. That's why the glass with the most water has the lowest sound.

This kind of idiophone is called a glass harmonica. Benjamin Franklin, a famous American inventor and politician, was so fascinated by it that he invented a musical instrument called a glass armonica. The armonica consisted of a long wooden box supported by four legs. Inside the box was a row of glass bowls, one inside the other, mounted on a long stick. When the bowls were turned by a floor pedal, their rims got wet from water at the bottom of the box. To play the armonica, the player rubbed his finger against the rotating wet rims. The glass armonica was so popular in the late 1700s that famous composers such as Beethoven and Mozart wrote music for it.

What happens if...

● you rub around the entire rim of the glass?

● you try other idiophones — slap the backs of two wooden spoons together, shake dried gourds, hit clay flower pots suspended by string from a branch, run your thumbnail across a comb?

SYNTHESIZERS

The synthesizer is a musical instrument developed in the twentieth century. It is played on a piano-like keyboard that is connected to a small computer. A musician can feed information into the computer that will produce certain types of sounds when he plays the keyboard. He might preset the synthesizer to imitate the sound of an instrument such as a clarinet or a flute. He might store several lines of music in the memory of the computer and play them back together or separately or at a different speed. He might enter the melody of a song into the computer and have it produce the chords and rhythm for that piece of music. Synthesizers are commonly used today to accompany singers and to produce recorded music in combination with non-electronic instruments.

AMPLIFY WITH AIR

MAKE A RESONATOR

Stringed instruments use a sound box
to amplify (make louder) the sounds of the strings.
But xylophones — instruments with parallel bars
of wood that are struck with mallets — simply use columns of air
to make the sounds loud enough for an audience to hear.
Here's how it works.

You'll need:
- *a can opener*
- *3 empty 540 mL (19 ounce) cans*
- *packaging tape*
- *a large container or pail*
- *water*
- *an A tuning fork (440 cycles per second)*

1 With an adult's help, remove both ends of the three cans.

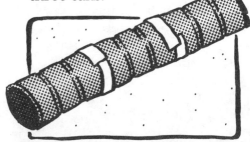

2 Tape the cans securely together to form one long tube.

3 Fill the pail with water.

68

4 Strike the tuning fork on a hard surface and hold it in the air. Can you hear it? Now do the same but hold the fork to your ear to hear its pitch more loudly.

5 Put the can-tube in the water. While holding the tube, strike the tuning fork on a hard surface to make it vibrate. Hold the fork directly above the tube horizontally. Move the tube and fork slowly up and down in the water until you hear the fork at its loudest.

? How does it work?

When the tuning fork is struck, it vibrates at a certain frequency (times per second). The vibrating fork makes the air in the tube vibrate. When the air column in the tube vibrates at exactly the same frequency as the tuning fork, they are in resonance and this makes the sound of the fork much louder.

NIGHT SOUNDS

What's one of the most annoying sounds that you can hear at night? Snoring. Then why do people do it? When you snore, you're likely breathing through your mouth instead of your nose. As you breathe in and out, different parts of your mouth and throat vibrate to make the snore sound. These parts might include your lips and cheeks, the roof of your mouth in the back of your throat, and the uvula, the little flap that hangs in the middle of your throat. Why do snores sound so loud at night? Because everything else is so quiet.

Xylophones work the same way to make notes louder. Underneath each bar on a xylophone is a tube of air closed at the bottom similar to the tube you made. These tubes are called resonators. Short bars, which have a higher pitch, have short resonators. The longer the bar, the longer the resonator needed to make the sound of the bar loud enough for an audience to hear.

PIANO AND DRUM OVERTONES

When you play a single note on an instrument, you hear a single sound or tone. Right? Actually, what you hear is a basic note combined with a few higher notes, or overtones, that blend together to give the one note that you hear. Here's how to find overtones with a piano and a drum.

Piano overtones

You'll need:
- *a piano*

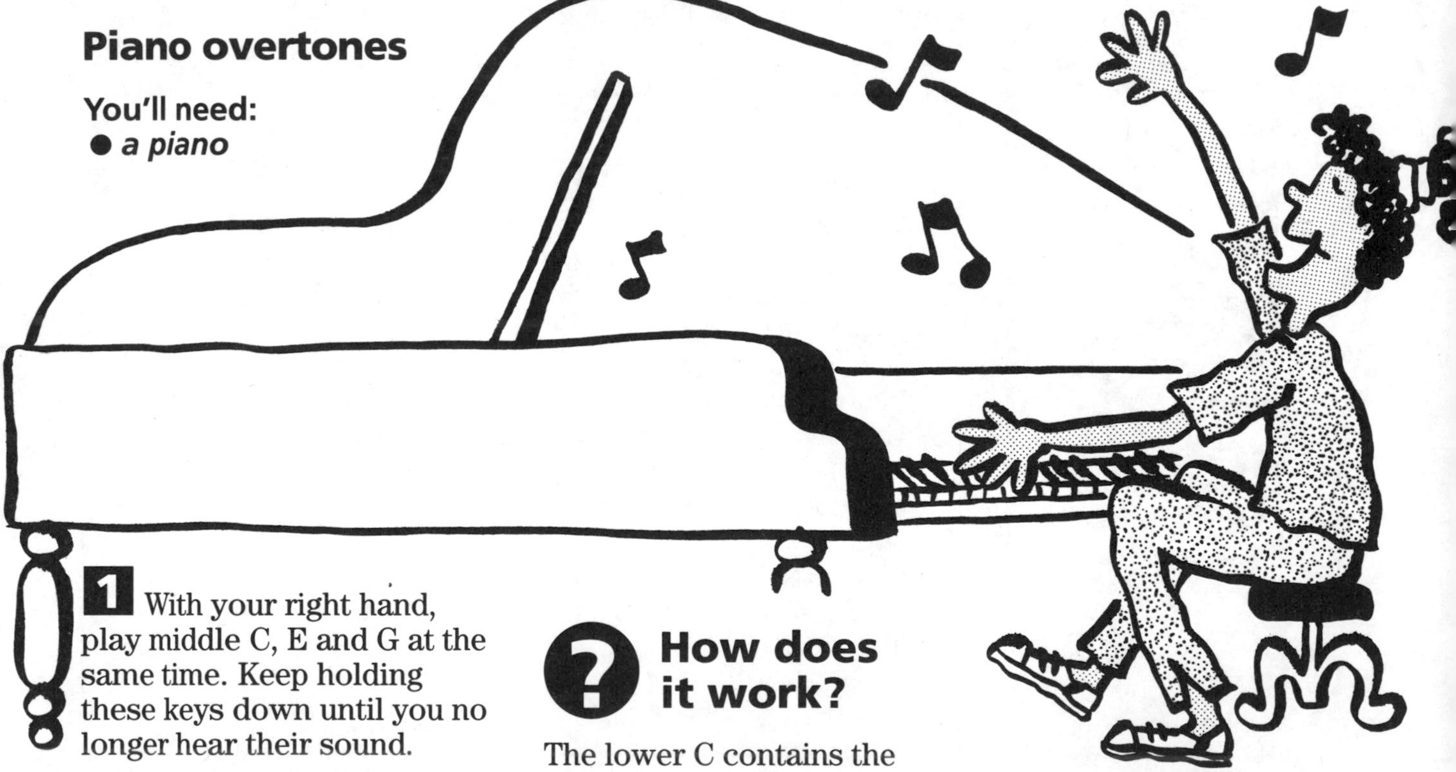

1 With your right hand, play middle C, E and G at the same time. Keep holding these keys down until you no longer hear their sound.

2 Still holding the three keys down, play a C two octaves lower loudly and sharply with your left hand. Can you hear the three original notes faintly?

? **How does it work?**

The lower C contains the overtones C, E and G as part of its sound. If you played the lower C by itself, you wouldn't hear the overtones. But since you are still holding down its overtones when you strike it, it causes the overtones to vibrate and make sound. Each note has its own overtones. What happens if you press G, B and D as overtones for a lower G? What other overtones can you find?

Drum overtones

You'll need:
- *a can opener*
- *2 empty 540 mL (19 ounce) cans*
- *1 empty 341 mL (12 ounce) can*
- *packaging tape*
- *scissors*
- *an old rubber glove*
- *a rubber band*

1 With an adult's help, remove the ends of the three cans and strip off their labels.

2 Join the three cans together with the packaging tape to make a single cylinder.

3 Cut the rubber glove in half. (You won't need the finger part.) Make one cut through the bottom half of the glove so that it opens up into a long rectangle.

4 Stretch the rubber rectangle tightly over one end of the cylinder and use the rubber band to hold it in place. Trim off the excess rubber.

5 Hold the drum between your knees. Hit the drumhead near its edge with the fingers of one hand flat and close together. Cup the other hand and hit the middle of the drumhead with the tips of the fingers. Do you hear two kinds of sounds? Use these sounds to create a rhythm.

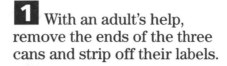

? How does it work?

When you hit the drum in the middle, the basic note of the drum vibrates more than the overtones, and so determines the kind of sound you hear. When you hit the drum near the edge, you allow more overtones to blend with the basic note and make a different kind of overall sound. This is a bit like a trampoline. If you stand next to a person who is jumping in the middle, you feel very strong vibrations. But if you stand near the edge of the trampoline, the vibrations feel weaker.

Drums that are played with sticks or mallets instead of hands vary their sounds in other ways. A snare drum has two membranes, or heads, made of plastic or skin. The bottom head has a set of tight wires called a snare stretched across it. When the drummer hits the top head, the snares vibrate against the bottom membrane giving a sharp, crisp sound. A bass drum also has two drumheads. It is much larger than a snare drum and is played by pressing a pedal connected to a felt-covered beater. It gives a short, deep thud.

CREATE SOUND EFFECTS

Films, radio shows and live plays use sound effects to make their stories realistic. Today, recordings of real objects and events are used for many of these sound effects. But if such recordings are not available, as they weren't in the past, many sound effects can be produced with objects around the house. Why not write a play and try them out with some friends?

1 Rain – Make a large cone out of waxed paper and hold it at an angle. Sprinkle salt from the top of the cone so that it runs down the inside. The salt hitting the waxed paper makes the rain sound.

2 Thunder – Hold a thin flexible cookie sheet by one corner and shake it.

3 Ocean waves against the shore – Put 50 mL (1/4 cup) dried peas into a large plastic bowl. Tilt the bowl slowly back and forth so that the peas slide along the bottom from side to side.

4 Fire – Crumple an empty cellophane bag that held rice or pasta.

5 Train – Turn the handle of a flour sifter back and forth so that each turn covers a half circle. Start slowly and gradually increase the speed.

6 Jet plane – Run a hair blower at low speed.

7 A voice on the telephone – Speak into a plastic cup.

8 Footsteps in the snow – Fill a small plastic bag with flour and tie it tightly. Hit it against a hard surface in the rhythm of footsteps in deep snow.

9 Foghorn – Blow across the top of an empty pop bottle.

WOODPECKER CODE

Woodpeckers have an unusual way of communicating. They drum or tap on a tree trunk or branch with their sharp beaks. Their drum rolls send a variety of messages such as: "This is a good place for a nest"; or "This is my territory"; or "I need a break from sitting on these eggs." Ornithologists (bird scientists) can identify different species of woodpeckers by listening to the length of the drum rolls and the pauses between them. For example, the black woodpecker has a drum roll of 38 to 43 beats, which are delivered three times per minute. The greater spotted woodpecker's drum roll is much shorter — 12 to 16 beats — and is repeated 8 to 16 times a minute.

SENDING SOUND

How many ways can you think of to send sound over long distances? You've probably thought of the telephone and the telegraph, but what about speaking tubes or bull-roarers? When you do the next experiments, you'll find out what these devices are and how you can use them to send sound. You'll also get a chance to set up your own phone system and build a telegraph for sending messages in Morse code.

TUBE TALK

CHANNELLING SOUND WAVES

Before modern intercoms were invented, people in large houses would communicate by talking through long tubes. Try this experiment to see how it worked.

You'll need:
- *a wristwatch that ticks*
- *5 or more cardboard paper-towel tubes*
- *scissors*

1 In a quiet room, hold the watch about 30 cm (12 inches) away from your ear. Notice how loud the ticking sounds.

2 Now hold one end of a paper-towel tube to your ear and hold the watch at the other end of the tube. Compare the sound you hear with Step 1.

3 Join two tubes together by making three 2 cm (3/4 inch) long slits at the end of one tube and slipping it inside the other tube.

4 Lay the watch on the floor. Put one end of the joined tubes over the watch. Hold the other end to your ear. How does the watch sound?

5 Join three, then four, then five tubes together. Listen to the watch each time you add a tube.

6 Add as many tubes as you can collect. Do you ever reach a length where you can no longer hear the watch?

❓ How does it work?

When you first listen to the watch without the tube, sound waves coming from the watch spread out into the room in all directions. When the sound waves travel from the watch through the tube, the walls of the tube stop the sound waves from spreading out and the sound is channelled directly to your ear.

In the 1800s people who lived in huge mansions would talk to their servants through speaking tubes. The tubes were metal pipes about the width of a toilet-paper roll and up to 30 m (100 feet) long. The pipes were inside the walls of the house; their ends came out of the wall and were flared like a funnel.

Captains of ships also used speaking tubes to give directions about speed changes to sailors in the engine room. The captain blew on a whistle inside the flared end of the tube to get a sailor's attention.

NOISY PIPES

Why do the water pipes in your home sometimes make strange sounds when you turn the faucet on and off? Some parts of the pipes are narrower than others. When water suddenly flows through these parts, the water is churned up and bubbles are formed. The air bubbles vibrate back and forth and their vibrations are amplified by the pipe and the surrounding walls.

THE BIGGEST BANG

How much noise do you think you can make with paper or cardboard? Try this experiment to find out. You're sure to get a bang out of it.

You'll need:
- *scissors*
- *a 22 cm (8 1/2 inch) square piece of thin cardboard (like a file folder)*
- *a 22 cm (8 1/2 inch) square piece of thicker bristol board*
- *a ruler*
- *tape*

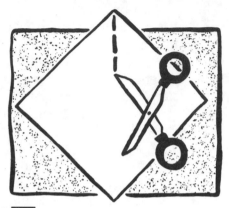

3 Lay the triangle piece on a table with the longest side closest to you.

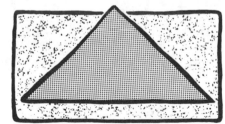

1 Cut the thin cardboard square in half on the diagonal. Discard one of the triangles.

2 With a blade of the scissors, score the bristol board square from one corner to the other on the diagonal.

4 Lay the square on top of the triangle so that the end of the scored line is 1 cm (1/2 inch) from the top point of the triangle (as shown).

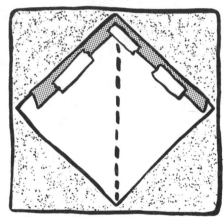

5 Fold the sides of the triangle over the edges of the square and tape them down.

6 Fold the square in half along the scored line so that the triangle folds in half inside it.

❓ How does it work?

When the triangle flicks outwards, it hits against the air with such a sudden force that it causes the air to vibrate. The vibrations are heard as a loud bang.

A similar thing happens during a thunderstorm. When lightning travels through the sky, it rapidly heats up the surrounding air. Since hot air expands (spreads out), the heated air pushes out against the adjacent air molecules causing them to vibrate. This push is so sudden and happens with such great force that it makes a noise that sounds like an explosion or thunder.

7 Hold the folded square at the untaped end. Raise it above your head and swing your arm down quickly.

8 Does the sound change if you use larger squares? What happens to the sound when you make the triangle out of other kinds of paper: waxed, bond, foil, construction, a paper bag?

HOW FAR AWAY IS THUNDER?

Does the crack of thunder sometimes sound like it's right over your head? Well, it may be farther away from you than you think. Since light travels faster than sound, you can use the flash of lightning you see during a thunderstorm to figure out how far away the thunder is. When you see lightning, count the number of seconds until you hear thunder. Divide the number of seconds by 3 to tell you how many kilometres away the lightning struck. (Divide by 5 to find the distance in miles.) This also tells you where the thunder is. So, if you count 15 seconds between the lightning and the thunder, the storm is 5 km (3 miles) away.

THE TALKING MACHINE

RE-INVENT THE PHONOGRAPH

When Thomas Alva Edison invented the phonograph in 1877, he made a machine that both recorded and played back the human voice. But the sounds that were heard were very weak. Try this experiment to find out how the sounds were amplified without the use of electricity.

You'll need:
- *a sheet of stiff paper 22 cm by 28 cm (8 1/2 by 11 inches)*
- *sticky tape*
- *2 sewing needles*
- *an old record (make sure it's a record that nobody wants any more because it won't be usable afterwards)*
- *a record player*

1 Put the record on the turntable and turn on the record player. Do not turn on the volume.

2 While the record is spinning, hold one of the sewing needles at an angle on the surface of the record. What do you hear?

3 Roll the paper into a cone with a wide mouth and use the tape to hold it together.

4 Stick the needle through the cone about 1.5 cm (1/2 inch) from the smaller end.

5 The sharp end of the needle should stick out about .5 cm (1/4 inch).

6 Now hold the cone so that its needle is resting gently on the rotating record. Put your ear to the open end of the cone. What do you hear? What do your fingers feel as they hold the cone? Compare this to the needle without the cone.

? How does it work?

The record has one very long groove that starts on the outside and winds round and round until it reaches the centre. Inside the groove there are many tiny wiggles. As the record spins, the needle follows this groove and vibrates back and forth as it hits the wiggles. The more wiggles, the faster the needle vibrates and the higher the sound. The vibrations of the needle by itself allow you to hear the sounds, but they are very weak. The needle on the cone makes the cone vibrate the air inside it. This allows you to hear a much louder sound than if the needle were vibrating by itself.

Edison's phonograph worked the same way to amplify sound. It also had a needle that vibrated in a groove, but the groove was on a rotating cylinder that was covered with tin foil. To make the sounds coming from the cylinder louder, a large horn similar to your paper cone was attached to the needle. About ten years later, another inventor replaced the cylinder with a flat disc similar to today's records. The horn continued to be used to amplify sound until the 1920s when electrical techniques started to be used.

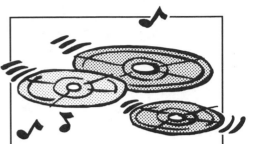

What happens if...

● you use your fingernail to make the sound?

● you use a smaller cone?

● you use a yogurt container instead of a cone?

COMPACT DISCS

The most up-to-date method of recording and listening to sound is on a compact disc. A compact disc is made out of a plastic material called polycarbonate. One side — the side where sound is recorded — is covered with aluminum. The surface of the aluminum has about 10 billion microscopic holes in it called pits. To give you an idea of the size of a pit, one hair from your head is as wide as 200 pits laid side by side. To play a compact disc, a laser light beam shines on the rotating disc. When the beam hits the flat parts — the land — between the pits, the light is reflected back to a small computer. The frequency of these reflections gives the computer information about what is on the disc. This information is converted into the sound you hear.

TRAIN LANGUAGE

MAKE A TRAIN WHISTLE

Train whistles are used to communicate messages to other trains, people and train stations over long distances. Here's a way to find out about the special language of train whistles.

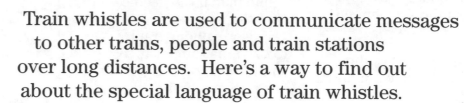

1 Cut the tube lengthwise in a straight line from one end to the other.

2 Roll the tube up to make a skinnier tube that measures 1.5 cm (1/2 inch) across each end. Use the tape to hold the roll together.

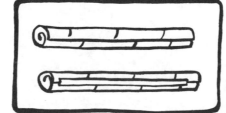

3 Cut a piece off the roll so that it measures 19 cm (7 1/2 inches) long.

4 Inflate the balloon, twist its neck and clamp it with the clothes pin so that no air escapes.

You'll need:
- *scissors*
- *a paper-towel cardboard tube*
- *sticky tape*
- *a balloon*
- *a clothes pin*

82

5 Stretch the neck of the balloon over one end of the tube.

6 Holding the balloon between your knees, remove the clothes pin and untwist the balloon's neck. Using your thumbs and first two fingers, stretch both sides of the neck open and closed.

7 Practise making short blasts and long blasts. Then try some of the train whistle messages in the box on this page.

❓ How does it work?

The air escaping from the balloon causes the stretched sides of the neck to vibrate. These vibrations travel into the tube, making the tube and the air inside it vibrate.

A train whistle works in the same way. The whistle actually has three tubes or flutes. Each flute has a thin brass disc that fits across one end but allows a small amount of air into the tube. Air that is under a lot of pressure is forced into the flutes just like air goes from the balloon into your tube. The air pressure makes the discs vibrate, which makes the sound we hear.

MOOSE ROMANCE

The pitch of train whistles is higher now than it was a few years ago. That's because moose used to think that the lower-pitched whistle was the love call of another moose and would run onto the tracks to see it. To stop the moose from running onto the tracks when a train was approaching, the pitch of the trains' whistles was changed.

TRAIN WHISTLES

An **o** means make a short blast.
A — means make a long blast.

Sound	Meaning
o	to another train: Stop!
— —	release brakes, go ahead
o o o (when standing)	the train is about to back up
o o (when standing)	the train is starting to move
— — —	the train is coming to a station or crossing at road level
o o o o o o o o	warning for people or animals to get off the tracks
o o o (when running)	stop at the next station
o o o o o (when running)	increase speed

SOAK UP SOUND

EXPLORE ACOUSTICS

When architects design large rooms such as auditoriums, they plan to use materials that absorb sound. Try this experiment to find out what materials are best for absorbing sound.

You'll need:
- a battery-run radio or an alarm clock
- ● a cardboard box that can be closed
- ○ towels
- ● a blanket or pieces of material
- ○ newspapers — crumpled or folded
- ○ carpet scraps
- ○ plastic bags
- ○ sweaters
- ○ a pillow

1 Turn the radio on to play music at a high volume, or turn on the alarm on your clock.

2 Put the radio inside the box and close it. Does the music sound as loud as before?

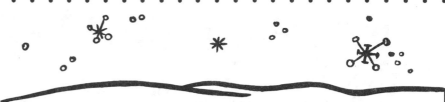

3 Find out which of the materials you've collected works best to reduce the loudness of the music. To do this, wrap the materials individually or in different combinations around the radio while it is playing inside the box.

4 Try the same experiment with the radio tuned to a speaking program. Are the same materials as effective as before?

❓ How does it work?

Materials with many small air spaces are best for absorbing sound. The air spaces trap and hold the sound. Offices, factories, hospitals, restaurants and auditoriums use carpeting, curtains and special acoustic ceiling tiles with tiny holes to cut down unwanted noise.

In large halls where music is performed, architects must be careful to use the right amount and kind of sound-absorbing materials. If too much sound is absorbed, musicians feel that their music sounds muffled. If not enough sound is absorbed, the sounds echo or bounce off the walls and the music is unclear or jumbled.

QUIET AFTER A SNOWFALL

Have you ever noticed how quiet it is just after a snowfall?

There are millions of tiny spaces inside and surrounding the flakes that make up the freshly fallen snow. These spaces absorb or trap sound. This may be an advantage in a large city since it temporarily cuts down on noise pollution. But on Antarctic expeditions, people who are more than 5 m (15 feet) apart in freshly dug snow tunnels must shout to be heard.

FASHION AND THE SOUND OF MUSIC

Sometimes it's not the fault of the architect if music doesn't sound right in an auditorium built for musical performances. In London, England, the Royal Albert Hall has been a popular place for concerts since 1871. But in the 1930s, people complained about the sound of the music performed there. Someone discovered that since women were no longer wearing long dresses made of several layers of material, not as much sound was being absorbed. This problem was solved by adding special fibre tiles to the walls.

WHIRL A BULL-ROARER

For thousands of years, bull-roarers have been used by various cultures in ceremonies. The sound made by the bull-roarer as it whirls through the air represents the voices of spirits and thunder. Here's your chance to make your own modern bull-roarer.

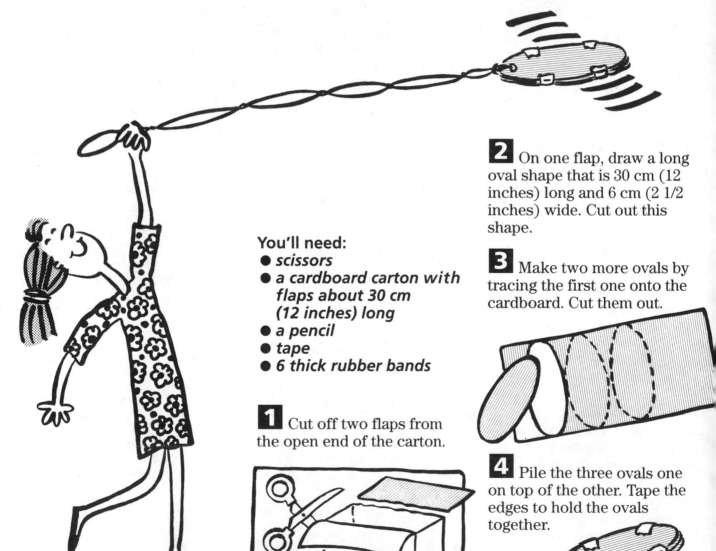

You'll need:
- *scissors*
- *a cardboard carton with flaps about 30 cm (12 inches) long*
- *a pencil*
- *tape*
- *6 thick rubber bands*

1 Cut off two flaps from the open end of the carton.

2 On one flap, draw a long oval shape that is 30 cm (12 inches) long and 6 cm (2 1/2 inches) wide. Cut out this shape.

3 Make two more ovals by tracing the first one onto the cardboard. Cut them out.

4 Pile the three ovals one on top of the other. Tape the edges to hold the ovals together.

5 With the point of the scissors, make a hole through all three layers 2 cm (3/4 inch) at one end of the oval.

6 Join the rubber bands together to make a long rubber chain. (To join one band to another, overlap the ends to make a small hole. Put the end of the underlying band through the hole and pull tightly.)

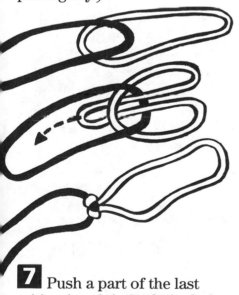

7 Push a part of the last rubber band through the hole in the cardboard. Bring the rest of the chain over the end of the oval, thread it through

the end of the rubber band and pull. The chain should now be firmly attached to the oval.

8 Whirl your bull-roarer in an open area away from people. To do this, hold the free end of the chain and rapidly whirl the bull-roarer around in a circle above your head.

❓ How does it work?

As you whirl the bull-roarer through the air, the oval or blade is spinning around rapidly. You can see this when you let the blade hang down immediately after you've stopped whirling it. The spinning blade pushes against the surrounding air molecules causing the air to vibrate. These vibrations travel to your ears as sound. The whirling blades of a helicopter make a noise for the same reason. Their noise is much louder than a bull-roarer's because the blades are larger and there is more than one rotating.

HOW DO RATTLESNAKES RATTLE?

When a rattlesnake senses danger, it uses its tail to make a distinctive rattling sound. Although the pitch of the rattle varies with the size of the snake, the sound is always made in the same way. The snake's tail is made of a stack of hollow, hard brown scales that look like tiny match boxes with rounded corners and edges. Since the scales are loosely connected, they rub against each other when the snake vibrates its tail. This makes the warning rattling sound.

FIDDLING WITH A PHONE

MAKE A TELEPHONE

It took Alexander Graham Bell several years of experimenting before he built a telephone that worked. It will take you a lot less time to put this phone together, but you'll have just as much fun fiddling around with it.

You'll need:
- *a hammer*
- *a nail*
- *2 empty 540 mL (19 ounce) cans*
- *heavy string*

1 With an adult's help, use the hammer and nail to punch a hole in the centre of the bottom of each can.

2 Cut a length of string 7 m (21 feet) long. Push one end of the string through the hole in one of the cans and make a large knot so that the string will not pull out. Do the same with the other end of the string and the other can.

3 With a friend, take your telephone outside. Hold the telephone so that the string is stretched tightly. Talk into your can while your friend holds her can to her ear.

4 Make the same telephone with strings of other lengths. Which length gives the clearest sound?

5 Now try a conference call. Get a third person to tie another string and can to the middle of the original string. How many lines can you add and still have a clear conversation?

? How does it work?

When you speak into the can, the sound of your voice makes the bottom of the can vibrate. The vibrations travel down the string (you can feel them if you touch the string gently) to the bottom of your friend's tin can. From there, the sound travels through the air in the can to her ear.

When you speak into a real telephone, a metal disc vibrates just as the bottom of the can did. The vibrating disc causes tiny grains of carbon (black particles that are the same material as the black part of your pencil) to press together and then move apart. These moving grains determine the amount of electricity that flows through them. The changing electric current flows through the telephone wires to the receiver of the other telephone. A tiny magnet in the receiver changes the electric current into sound vibrations that we can hear.

What happens if...

● you use smaller or larger cans?

● you use plastic containers instead of cans?

● you let the string touch something while you talk?

● you let the string hang loose?

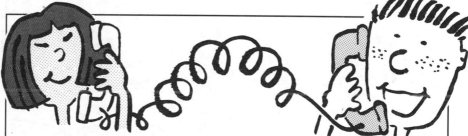

BROKEN TELEPHONE WITH A TWIST

The next time you and your friends play Broken Telephone, try sending a tongue twister along the line. Since most tongue twisters are familiar, make up your own. Just start each word with the same letter.

For example: Dreadful dragons devoured Dracula's delicious Danish doughnuts. You might not recognize the final message, but you'll all have a good laugh when you hear it.

SEND IT IN CODE

Once the telegraph was invented by Samuel Morse in 1843, people were able to communicate over long distances through electrical wires. Since only clicks were transmitted over the wires, Morse used these clicks to create a code of dots and dashes that stood for the letters of the alphabet and numerals. This system is called the Morse code. Why not make your own telegraph and have a coded conversation with a friend?

Important: Do not make substitutions for these materials.

You'll need:

scissors
4.5 m (15 feet) of single-strand bell wire
2 blocks of wood, each about 18 cm (7 inches) by 15 cm (6 inches) by 2.5 cm (1 inch)
a hammer
5 carpet tacks
electrical tape
a new "D" battery (1.5 volts)

- *an iron nail 7.5 cm (3 inches) (If a nail is attracted to a magnet, it has iron in it.)*
- *an empty juice can 1.36 L (48 ounces)*
- *a can opener*

1 Cut off a 15 cm (6 inch) piece of wire from the full length. Use the scissors to strip off 1 cm (1/2 inch) of plastic covering (insulation) from one end of the wire and 2.5 cm (1 inch) of insulation from the other end.

2 Bend the 2.5 cm (1 inch) of exposed metal to make a hook.

3 Attach this wire to a wood block by hammering a carpet tack into the centre of the block at the midpoint of the wire. The head of the tack will keep the wire in place.

4 Tape the battery to the block near the end of the wire with the 1 cm (1/2 inch) stripped off. Tape the end of the wire securely to one end of the battery.

5 Starting at the midpoint of the long wire, wrap the wire neatly around the nail so that the loops are placed immediately next to each other. The wire should not overlap. The whole nail should be covered except for a bit near the tip. You've just made an electromagnet (a magnet that works by electricity).

6 Hammer the nail into the other wood block about 5 cm (2 inches) from the end of the block.

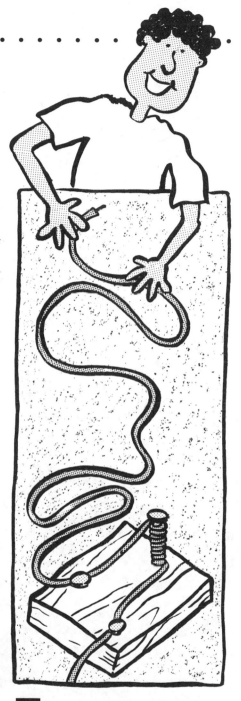

7 Fasten the two wires leading from the nail to the wood with carpet tacks, as shown. Strip 1 cm (1/2 inch) of insulation off the end of each wire.

10 Ask an adult to help you with this step. Remove the bottom of the juice can with the can opener. Cut a strip of metal 2.5 cm (1 inch) wide from the bottom. Tape all sharp edges.

8 Tape the end of one wire securely to the other end of the battery.

9 Attach the stripped end of the other wire with a carpet tack to the other wooden block just beneath the hooked end of the short wire.

11 Bend the metal strip to make a "Z" shape with a straight back. With a tack, attach the bottom of the "Z" to the end of the block behind the electromagnet. The top of the "Z" should be almost touching the nail head.

12 To make your telegraph work, hold the short wire by the insulation and touch the hook to the tack underneath it. You will hear a click each time you do this. Use the code below to tap out a message. If you want to send a message to a friend in another room, double the length of wire you buy, and put the block with the metal "Z" in your friend's room. Keep the block with the hook in your room. If your friend wants to answer, she must have a similar set.

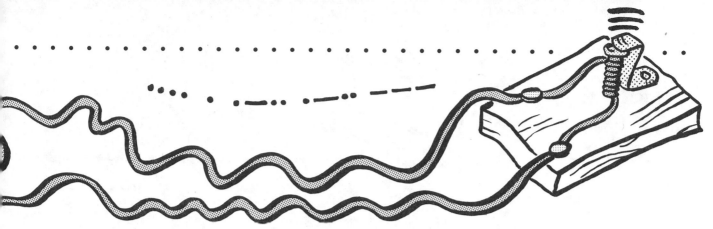

TROUBLESHOOTING:

If your telegraph doesn't work the first time you try it, check these items in this order:

1. Are the wires attached securely to the ends of the battery?
2. The top of the metal strip might be too high above the nail.
3. Your battery might be too weak. Try two in a column as in a flashlight.

❓ How does it work?

When you touch the hook (key) to the tack, you complete an electrical circuit. Electricity travels in an unbroken circle or circuit from the battery through the wires to the electromagnet. This enables the electromagnet to attract the metal strip to it. That's when you hear a clicking sound. When you lift the key, the electrical circuit is broken; the metal is no longer attracted to the electromagnet and lifts up.

International Morse Code:
A dot is a click followed by a very short pause and a dash is a click followed by a longer pause.

A	• —	P	• — — •	1	• — — — —
B	— • • •	Q	— — • —	2	• • — — —
C	— • — •	R	• — •	3	• • • — —
D	— • •	S	• • •	4	• • • • —
E	•	T	—	5	• • • • •
F	• • — •	U	• • —	6	— • • • •
G	— — •	V	• • • —	7	— — • • •
H	• • • •	W	• — —	8	— — — • •
I	• •	X	— • • —	9	— — — — •
J	• — — —	Y	— • — —	0	— — — — —
K	— • —	Z	— — • •		
L	• — • •			period	• — • — • —
M	— —			start	— • —
N	— •			end	• — • — •
O	— — —			error	• • • • • • • •

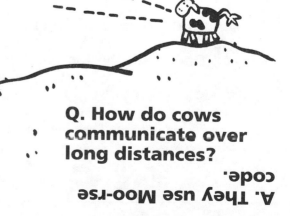

Q. How do cows communicate over long distances?

A. They use Moo-rse code.

93

MATCH IT!

These common sound sayings and their meanings are all mixed up.
Can you match the meaning with the correct saying?

Saying

1 You could hear a pin drop.

2 In one ear and out the other.

3 Actions speak louder than words.

4 She spoke off the cuff.

5 He talks my ear off.

6 You took the words right out of my mouth.

7 Don't blow your own horn.

8 The cat's got his tongue.

9 Mark my words.

10 Your advice is falling on deaf ears.

Meaning

A Don't boast or brag.

B We are thinking exactly alike.

C It was suddenly very quiet.

D He isn't speaking.

E Remember what I say because it will probably come true.

F She spoke without preparing a speech.

G Heard but not remembered.

H He is ignoring your advice.

I He talks too much.

J What you do is more important than what you say.

Answers: 1-C, 2-G, 3-J, 4-F, 5-I, 6-B, 7-A, 8-D, 9-E, 10-H

GLOSSARY

Acoustics The science of sound

Amplifier A device that makes sound louder

Eardrum A thin membrane across the middle ear that vibrates

Echo A repeating of sound after it is reflected off a hard surface

Echolocation A method of finding the distance and position of an object by measuring the time it takes sound or radio waves to bounce back from the object

Frequency The number of times an object vibrates in one second. The greater the frequency of a vibrating object, the higher the pitch of its sound. Frequency is measured in hertz.

Hertz A unit of measurement that is used to express the frequency or pitch of sound

Infrasound Sound that is too low for the human ear to hear

Idiophone An instrument made from a material that produces a sound from itself. Bells, cymbals, gongs and maracas are all idiophones

Overtone A higher tone that combines with the main tone of a note played on a musical instrument

Pitch The highness or lowness of a musical sound

Radar A device that determines the distance, direction and speed of an unseen object by the reflection of radio waves. The letters stand for ra(dio) d(etection) a(nd) r(anging).

Resonance When a vibrating object causes another object with the same natural frequency to vibrate, the two objects are in resonance

Sonar A device that locates objects by sending out and receiving back reflected sound waves. The letters stand for so(und) na(vigation) a(nd) r(anging)

Sound box A hollow box that is placed behind something that makes sound (e.g. strings) in order to make that sound louder

Sound wave Sound travels from its source in sound waves that are produced when an object vibrates.

Supersonic Able to move at a speed greater than the speed of sound

Synthesizer An instrument that changes electrical impulses into sound. It is operated from a keyboard.

Ultrasound A sound above a frequency of 20 000 vibrations per second. This sound is too high to be heard by the human ear.

Vibration The regular back and forth movement of an object

INDEX